International Yearbook of Oral History and Life Stories

VOLUME I

1992

International Yearbook of Oral History and Life Stories

VOLUME I

Memory and Totalitarianism

Special Editor

LUISA PASSERINI

OXFORD UNIVERSITY PRESS
1992

Oxford University Press, Walton Street, Oxford OX2 6DP
Oxford New York Toronto
Delhi Bombay Calcutta Madras Karachi
Petaling Jaya Singapore Hong Kong Tokyo
Nairobi Dar es Salaam Cape Town
Melbourne Auckland
and associated companies in
Berlin Ibadan

Oxford is a trade mark of Oxford University Press

Published in the United States
by Oxford University Press, New York

British Library Cataloguing in Publication Data
Data available

Library of Congress Cataloging in Publication Data
Data available
ISBN 0–19–820248–2

Typeset by Cambrian Typesetters, Frimley, Surrey
Printed and bound in
Great Britain by Biddles Ltd.,
Guildford and King's Lynn

Preface

The publication of the first *Yearbook* marks a great step forward for all of us in the humanities and social sciences who seek to understand the recent past and the changing present through autobiographical testimony, whether in the form of written autobiography, oral history, or life-story interviews.

Each year we shall be publishing a new *Yearbook* which like this, the first, stands in its own right as an independent, coherent book on a single theme, but also contributes cumulatively to our aims as a regular annual publication.

The series is both international and interdisciplinary. It builds on the experience and co-operation developed through the publication of *Life Stories/Récits de vie* in Europe and the *International Journal of Oral History* in North America and was conceived as a joint successor to both journals. The editorial team is international. It combines a strong European network with a parallel North American team, and also links to Australasia, Latin America, and Eastern Europe. For the third volume editorial responsibility will cross the Atlantic. We are committed to developing real, practical editorial and intellectual co-operation between cultures, nationalities, and continents. And between languages. Our earlier experience of publishing a two-language annual proved disappointing in terms of communication across the language barriers through this approach. So this time we publish in English only, but put much more of our efforts into translation. This first book includes articles which originated in German, Russian, and Italian and reviews of literature in these languages as well as in French and Spanish.

We are at the same time fundamentally interdisciplinary. The series is intended not only for contemporary historians, sociologists, and anthropologists but also for those with overlapping interests in problems of memory, autobiography, and narrative in psychology and literature. Our editorial team includes recognized leaders in all these fields.

This first *Yearbook* focuses on a theme which, in geographical terms, is much narrower than those which are to follow. The second is to be on cultural transmission between generations, and the third on migration and identity: two themes which are truly international.

The achievement of this volume is nevertheless close to the heart of our intentions. It crosses an ideological frontier which divided the world for most of our lifetimes. It wrestles with the most painful memories which we in Europe have of this century at the end of the Cold War. Yet, as we have manifestly seen since 1989, the political implications of the memory of European Nazism and Communism concern the entire world. The struggle for that memory and its meaning will shape our future: together, or apart.

Paul Thompson
October 1991

Contents

List of Contributors ix

1. Introduction 1
LUISA PASSERINI

2. Antagonistic Memories: The Post-War Survival and Alienation of Jews and Germans 21
FRANK STERN

3. Where Were *You* on 17 June? A Niche in Memory 45
LUTZ NIETHAMMER

4. A German Generation of Reconstruction: The Children of the Weimar Republic in the GDR 71
DOROTHEE WIERLING

5. After Glasnost: Oral History in the Soviet Union 89
DARIA KHUBOVA, ANDREI IVANKIEV, and TONIA SHAROVA

6. The Gulag in Memory 103
IRINA SHERBAKOVA

7. The Abduction of Imre Nagy and his Group: The 'Rashomon' Effect 117
ANDRÁS KOVÁCS

8. Mujeres Libres: The Preservation of Memory under the Politics of Repression in Spain 125
MARTHA ACKELSBERG

9. A Shattered Silence: The Life Stories of Survivors of the Jewish Proletariat of Amsterdam 145
SELMA LEYDESDORFF

10. Don't Forget: Fragments of a Negative Tradition 165
RENATE SIEBERT

REVIEW ARTICLES

Oral History and Italian Fascism 179
ALFREDO MARTINI

Stories of Everyday Life in Vichy France 184
DOMINIQUE VEILLON AND DANIÈLE VOLDMAN

Voices from the Choir: Reflections on the Development of Oral History in Russia 188
IRINA SHERBAKOVA

Aspects of Recent Oral History in Germany 192
ALEXANDER VON PLATO

BOOK REVIEWS

Michael Pollak, *L'Expérience concentrationnaire: Essai sur le maintien de l'identité sociale* (Luisa Passerini) 197

Alessandro Portelli, *The Death of Luigi Trastulli and Other Stories: Form and Meaning in Oral History* (Carl Ryant) 199

Marcella Filippa, *Avrei capovolto le montagne: Giorgina Levi in Bolivia, 1939–46* (Alessandro Portelli) 200

Joanne Rappaport, *The Politics of Memory: Native Historical Interpretation in the Colombian Andes* (Elizabeth Tonkin) 201

Lawrence Goodwyn, *Breaking the Barrier: The Rise of Solidarity in Poland* (John J. Fox) 202

Teresa Toranska (ed.), *'Them': Stalin's Polish Puppets* (Bronislaw Misztal) 204

Alain Brossat, Sonia Combe, Jean-Yves Potel, and Jean-Charles Szurek (eds.), *A l'Est la mémoire retrouvée* (Daniel Bertaux) 206

List of Contributors

MARTHA ACKELSBERG teaches politics and history at Smith College, Massachusetts. Her writings include *Free Women of Spain*.

ANDREI IVANKIEV, *see* Daria Khubova.

DARIA KHUBOVA, ANDREI IVANKIEV, and TONIA SHAROVA worked together as members of the Oral History Club while students of history at the Moscow State Institute for History and Archives. Tonia Sharova now teaches medieval history at the institute; Daria Khubova directs its Oral History Centre; and Andrei Ivankiev runs an independent translating and transcribing firm.

ANDRÁS KOVÁCS, after many years of oppression during which he earned money as a life-story interviewer, now teaches oral history at Budapest University.

SELMA LEYDESDORFF is Professor of Women's History at Amsterdam University. Her writings on Dutch and Jewish history include *We Lived as Human Beings*.

ALFREDO MARTINI teaches history at the University of Rome.

LUTZ NIETHAMMER is Professor of History at the Kulturwissenschaftliches Institut, Essen. Before his present research on East Germany he led the project on 'Life Stories and Social Culture in the Ruhr 1930–60', and his books include *'Die Jahre weisman nicht, wo man die heute hinsetzen soll'*, *'Hinterher, weiss man, dass es richtig war, dass es schief gegangen ist'*, and *'Wir kriegen jetz andere Zeiten'* from this research. He is co-editor of *Die volkseigene Erfahrung*, and an editor of *Bios*.

LUISA PASSERINI is Professor in the Methodology of Historical Research at the University of Turin. She was founder-editor of *Fonti orali* and her books include *Fascism in Popular Memory* and *Mussolini immaginario*.

TONIA SHAROVA, *see* Daria Khubova.

IRINA SHERBAKOVA is a freelance Moscow journalist who was among the first to record recollections of Stalin's concentration camps.

RENATE SIEBERT teaches history at the University of Calabria.

FRANK STERN teaches and writes on German history at the University of Tel Aviv.

DOMINIQUE VEILLON and DANIELE VOLDMAN are researchers at the Institut de l'Histoire du Temps Présent, Paris.

ALEXANDER VON PLATO teaches history at the University of Hagen and researched on the Ruhr and East German projects. He is co-editor of *'Wir kriegen jetz andere Zeiten'* and *Die volkseigene Erfahrung*, and an editor of *Bios*.

DOROTHEE WIERLING worked with Lutz Niethammer on the Ruhr and East German research projects. She is co-editor of *Die volkseigene Erfahrung*.

Book Reviews: Daniel Bertaux, EHESS, Paris; John J. Fox, Salem State College, Massachusetts; Bronislaw Misztal, Purdue University, Indiana; Alessandro Portelli, Rome University; Carl Ryant, University of Louisville, Kentucky; Elizabeth Tonkin, Queen's University, Belfast.

1

Introduction

LUISA PASSERINI

> Ich bin der Engel der Verzweiflung. Mit meinen Händen teile ich den Rausch aus, die Betäubung, das Vergessen . . . Meine Rede ist das Schweigen, mein Gesang der Schrei.
>
> (Heiner Müller, *Der Auftrag*; words used for the composition *Frau/Stimme*, by Wolfgang Rihm)[1]
>
> I am the angel of despair. With my hands, I give you the narcotic, the anaesthetic, the oblivion . . . My discourse is silence; my song a scream.

In the play *Memoria*,[2] written by Else Marie Laukvik and directed for the Odin Teatret by Eugenio Barba, two stories of people who survived the concentration camps are told. The first tells of young Moshe, who, almost dying of cold, is standing in the snow in Mauthausen with other naked prisoners, answering an interminable roll-call. But a memory comes to him, a memory of his rabbi and of a song he used to sing. That memory saves Moshe, who dances the old song until he can break the ice that imprisons his feet, and defeat cold and death. The second story is one of little Stella, witnessing the Nazis' assassination of her relatives and friends in her village in Galicia. She hides in a hole in the ground, where she remains for eight months. In that hole, among rats, her memory freezes in a desire for vengeance; but when liberation comes and she sees a public execution of Nazi officers, that vengeance appears incapable of real memory: vengeance does not reach back into the past, it remains chained to the uncontrolled screams of the excited crowd. Brought up short, Stella draws back, and in doing so perceives that only by distancing herself from the present can she retrace her true memory and her true self.

The two stories are taken from *Hasidic Tales of the Holocaust* by Yaffa Eliach, and in the play they are accompanied by beautiful Yiddish folk-songs. They remind us of two triumphs of memory: triumph over torture and over a 'wrong' type of memory, a frozen and rancorous one, in favour of life and innovation. But the play does not finish with this happy ending. It continues with a second part, where the main actress mechanically repeats pieces of memory—phrases, words, all jumbled up in an incoherent stammering—in which one can only distinguish some sense if one already knows the complete story: dreadful details of torture are mixed up with happy associations of times long gone by. And in the final ending, the two protagonists present the photographs of Primo Levi and Jean Améry (Hans Mayer),[3] who both committed suicide many years after the concentration camps had ceased to exist. Their images challenge us to have a less optimistic view of memory, enigmatically posing the question of our time and its relation to the recent past of Europe.

What are some possible meanings of the enigma posed by the play? A first conclusion is that today it is not sufficient to survive physically. A second is that not even sheer remembering is quite up to the tasks of the present. What is required is indeed not a simple and spontaneous memory, not the one that stems from a need for vengeance, for instance, but a memory of a memory, a memory that is possible because it evokes another memory. We can remember only thanks to the fact that somebody has remembered before us, that other people in the past have challenged death and terror on the basis of their memory. Remembering has to be conceived as a highly inter-subjective relationship.[4]

In the first part of the play, memory struggles with a totalitarianism that wants to cancel a whole people and their culture from the face of the earth; in this *corps à corps* it finds strength in its own lability: it associates the great and the small, the important aspects and the details, lightly, asystematically, simply out of love for pleasure and life. In its very essence, memory repeats and replays, reproducing the stories of ancestors and old masters, their songs, their traditions. But the painful faces of Jean Améry and Primo Levi remind us of the destiny of memory in our contemporary world; their images break up the thread of the story as well as the thread of memory. Like the laughter of the crowd watching the execution of the war criminals—a crowd that has reverted to barbarity because it is not capable of fully remembering the past—the mere repetition of what is remembered by, or said to, those who do not really wish to remember, itself also sounds like a broken and stammering gramophone: one word here and one there, a mixture of associations that have lost their original meaning, a language of folly and of non-communication. There is nothing left to transmit if nobody is there to receive the message.

The richness of the original stories about Moshe and Stella is lost in a disorder which can be compared to the excess of daily messages that we receive today, whereby not silence but shouting and loud noise circumscribe and confuse the human word. In the play, the final image is that of a stranger singing and crying under a tree, and this image expresses our condition: a memory that is a stranger to the world, which will not be able to become a house for human beings without accepting that stranger and his estrangement. Those who committed suicide left, by their gesture, a message for us: now it is our turn to remember and sing again the old song of the rabbi, but sing it in a way that is up to date with our time, yet contrasts with this time's tendency to forget. After all, memory is the tool we have in order to give meaning to our lives, if we understand it in the sense of an inter-subjective (or inter-human) word that connects different generations, times, and places.

This *Yearbook*, with which we start a series of annual books about memory and life experiences, tries to give some responses to the questions posed by the Odin Teatret in *Memoria*. It is not merely chance that led us to start this new publication with the theme 'Memory and Totalitarianism'. Europe's past has been brought up to date, and the question of understanding it has been made more urgent by the events of August 1991 in Moscow and in what used to be called the Soviet Union. The invasion of the streets by Russian people rejecting the *coup d'état* was the culmination of a process that had been initiated years before; once again it raises crucial questions in a way that cannot be eluded: To what extent can these events be considered the end of an era stretching from the First World War to the 1980s, when Europe experienced many forms of dictatorship? To what extent can these forms be grouped together? Can any sort of affinity be established beween them?

We are aware that the very title of this *Yearbook* runs the risk of equating all sorts of regimes under the label of totalitarianism. Such a 'reductionist' approach might be coupled with another possible impoverishment if we believed that the problem of memory could be reduced to a question of oral history or a collection of oral memories from living people. But the issue of memory today is such a massive one, involving much more than orality and individuals, that it would be absurd to think that we can do more than give our specific contribution to a much larger problem. What we shall try to do, therefore is to situate the findings of oral history in the context of contemporary memory, making some references to other fields, like the social sciences, but also to literature, music, the theatre, and the local toponymy. Thus, we can reasonably expect to avoid the second reduction we have mentioned; as for the first one, however, we ourselves cannot easily be reassured.

Let us put our stand in context. We speak from a place—Europe—which likes to portray itself as a laboratory for democracy, where the ability to speak (*l'aptitude à la parole*) appears to be the essential force of opposition. To be a democrat means to consider that the main political task is to acknowledge that every person has the right and should have the opportunity to speak and to engage in a real dialogue. We know all too well that our present means of expression very rarely serves the purposes of a real democracy, 'yet to be created', while ours is 'chronically suffocated',[5] and that we lack a real 'public word', a real possibility of the expression of free opinion in the public sphere. Therefore we speak from a contradiction, and yet this is our only strength: we are under no illusion that we hold the truth; we know that democracy's one truth lies in its rules, intended to let everyone express his/her truth. This is the European legacy, the main contribution that this part of the world has made to the world at large, although it must be added that until now democracy has been a very incomplete and imperfect legacy. On a historical level, there is much that would deserve further study in order to clarify the pluralistic nature of democracy in the European heritage. Suffice it to think of the multiple traditions of resistance to Fascism and Nazism: the roots of antitotalitarianism range from Communism to Liberalism in this case, and some of the most interesting areas for us today are represented by those minority groups that proposed hybrids like Liberal-Socialism, while trying to amalgamate the two contradictory ideals of justice and freedom.

On a theoretical level, we could say that, on the one hand, the concept of democracy emerging in Europe in the last two decades stresses formalism as an essential feature: 'democracy, as counterposed to all forms of autocratic government, is characterized by sets of rules that establish *who* is authorized to take collective decisions and by *which* procedures'. On the other hand, increasing attention is being given to the actual situations that allow the rules to be enacted: 'those who are called to decide, or to elect those who will decide, must be confronted with real alternatives, and able to choose between them'.[6] In practice, the preconditions for the exercise of democracy include full respect for ethnic and gender differences. Contributions from movements based on such differences indicate that democracy must be pluralistic and must include various forms, such as the combination of direct and representative democracies; this is the only way to revive participation by people in democratic institutions.[7] A major problem in theoretical and political terms would be for a 'left'—if such a word were to take up some new meanings, given the collapse of the old ones—to reformulate the concept of democracy. The task would involve distinguishing the historical form it took through the Anglo-American heritage, and therefore its connection with imperialism, from its potential strength in terms of rights, focusing on the needs of the majority of people, until now excluded from the actual exercise of democratic rights.

The main task for the immediate future is to translate the principles of formal democracy into something meaningful for all the people of the south and east of the world (if we can still adopt the old idea of Europe, a western area closely linked with the United States, as a geographical point of reference). If we confront our contradictory situation with that of the regimes which are generically labelled as totalitarian, we find that it differentiates itself from them, because they are obliged to possess a truth, whether it is placed in the laws of history, as in Marxism-Leninism, or in biology and race, as in Hitlerism, or the State, as in Fascism.[8] It thus becomes clear that, in our context, the use of the word 'totalitarianism' refers in the first place to the speaking or writing subject, i.e. ourselves. It is a subjective point of departure, which we assume will change during our research. We recognize that the dichotomy of despotism and free society has been essential to the self-awareness of Western societies, but at the same time we are aware of the elements of ideology in this dichotomy, implicit in the dialectics of enlightenment that originated it, involving at the same time both elements of domination and also of liberation.

It should be clear that the use of 'totalitarianism' on our part implies only a partial acceptance of its traditional implications. One of the most important ones is its link with the concept of 'masses', as containers that absolute powers fill with whatever contents they decide. It is a central assumption of most oral history that people always have something to say on what is proposed to or even imposed on them, or at least that potentially every individual has an understanding and an interpretation of his/her history as well as of History. From this perspective, we are all in some way the 'masses', in as far as we have all been subjected to the effects of uniformizing modernity, and yet, in certain respects, we all believe ourselves to be the subjects of our own lives, capable of counteracting those levelling influences. At the same time, we might see a powerful potential for equality in these forces, so long as we do not forget that any real equality must be based on the recognition and respect of differences.

As oral historians we use the term 'totalitarianism' in a partial and critical sense, but above all in order to remember that there have been similarities of oppression among systems of thought and power that were in many ways very different. In addition to this, there were even similarities and connections in the forms of subjectivity that Fascism and anti-Fascism fostered. This difficult theme has hardly been addressed by historians, but has not escaped the acute glance of literature. In the first volume of Peter Weiss's novel *Aesthetics of Resistance*, the protagonist, enrolled in the International Brigades fighting in Spain, expresses the roots of those similarities in a lucid self-analysis:

> associated with the desire for a fundamental change, for the construction of a new existence, there was the feeling of our affinity with the country where the domination of capital had been overturned and where workers' power had been constructed. Our indignation and our rebellion would have been hopeless, had this country not represented for us something indestructible . . . Our own despair allowed us to understand that there too outbursts of folly and anger could take place. We approved of the intolerance with which they acted over there. Any wait-and-see policy had to be avoided. A reconciliation, an arrangement were inconceivable. One could talk about aberration, mistakes, panic, but for us any blow, any violence were justified. That country was isolated, as we were isolated and in that isolation we were linked to each other. That link gave us the only possible means for persevering, and it was in such perseverance that the peerless, seducing images of October lived on. No doubt, no hesitation could trouble those images.[9]

One future direction for historians should be the investigation of the nature of those images, their influence and roles, and their changes over time. Oral history could contribute some of the tools necessary to analyse memory for these purposes.

To outline our starting-point more precisely, we must emphasize that we use the term 'totalitarianism' almost exclusively for its implications in the field of subjectivity, that is the historical study of memory, mentality, language, and cultural expression. Therefore we are interested in a very wide range of meanings of 'totalitarianism', extending to aspects that are of particular relevance to us, such as the idea of totalitarian mentality. This idea is usually based on a cult of consensus and authority that can be found in many circumstances, often within democratic regimes. Therefore, although this introduction is concerned with the European experience of totalitarianism, the theme calls for an expansion to include many other experiences that relate to totalitarianism in mentalities. One only needs to think about certain issues that are now being researched in oral history, such as racist attitudes against blacks in the South of the United States of America; or the cultural and psychological effects of apartheid in South Africa;[10] or the Aborigines' claim to their own history and to a new idea of history, in Australia.[11] The results from those researches, and the comparison between them and others, could lead to larger and more meaningful comparative studies of memory and totalitarianism in the future.

From the perspective of continuity between capitalist democracy, that is a democracy which has been historically inseparable from the market, and totalitarianism—an approach not only of the Frankfurt School of Thought, but of many anti-Fascist historians—we can accept two ideas: that totalitarianism is not only external to us but also inside ourselves, with its roots continuously present in our societies and our lives; that totalitarian systems are social systems like other ones, in the sense that their language and discourse have a meaning for their protagonists, even if that meaning is unacceptable to us.[12] If we see the connection between democracies and totalitarian regimes—both historically (in 1920, 28 European states could be described as democracies and by the end of 1938, 16 had succumbed to dictatorship)[13] and theoretically (the latter are transformations of the former and could develop only from democratic systems)—we can also see the conventionality of uniting a number of different regimes under the same title. Totalitarianisms are products of the twentieth century that go far beyond earlier manifestations of absolutism and autocracy in their effort to completely control political, social, and intellectual life, made possible by modern industrialism and technology. They also need to pretend to be democratic: that is why they not only require a system of propaganda as well as control of expression and communication, but must also impose on the whole population the necessity of reciting what 'has to be said'. Totalitarianisms need the comedy of unanimity, and that is why they consider every invention and creativity, including poetry and jokes, as dangerous, while they have to encourage the capacity for copying and repeating. Democracies, on the other hand, host structures that facilitate the latent application of force as the oppression of persons by persons through unperceived structural violence that shapes ideologies, values, and dependencies.[14]

Yet we also know that Communist and Fascist regimes adopted quite distinct approaches precisely in terms of techniques of terror and propaganda, party organization, and one-man leadership, which are especially relevant for subjectivity and therefore for memory. Only very recently have we reached a position to fully appreciate these differences, with the new 'fever of history' and the 'almost pathological interest in the past' that started in the Soviet Union after 1987,[15] and that is expressed in the very important work that individuals and groups like 'Memorial' have been doing (see Ch. 6). Their experience, now under way, sheds light on important aspects of memory, such as its capacity to permeate society in many ways: let us think, for instance, of the role that literature played in Eastern Europe and Russia, in recomposing and keeping alive collective memory in a society that had been forced into amnesia and had been subjected to a repeated rewriting of history, including the glorification of the tsars in the interest of Stalinism. Let us think of the changes of names of towns, roads and places that resulted in a cancellation of history as well as of geography.[16] The obsessive repetition of new toponyms, celebrating the revolution, tended to abolish all regional and cultural diversities, reducing the whole country to the same image, that of the centre: ideology replaced memory, not simply by cancelling it, but by trying to impose an artificial and homologizing memory based on an imaginary, non-existent Soviet people. And in considering daily life, let us think of the fact that false memories were forged, individuals were pushed to destroy their own letters and diaries, and an atmosphere of lies and falsehood was perpetuated. All this makes clear in just how many ways memory can be disputed: in symbols, statues, museums, archives, and names. If the series of studies published by Pierre Nora on *Les lieux de la mémoire*[17] had already shown to what extent collective memory and identity were connected with 'places'—in the largest sense of the word—recent developments in Eastern Europe have not only confirmed but widened our appreciation of the relevance of the 'theatres of memory'.[18]

At the same time, all this deeply preoccupies whoever deals with memory, for two reasons. The first is related to the end of 'the hope principle' (*le principe espérance*[19]), with the collapse of 'avowed socialism', which is not only the failure of a system, but also the sign of a deadly crisis of hope and prophecy throughout the whole of Europe. It was because in the 1920s something was lost—the millenarist hope of putting into practice a free society—that it was necessary to create the myth of a socialist paradise: it became necessary to create a legendary USSR and to propagate its legend.[20] When the promise of a better future, more beautiful and more just, was revitalized by 1968, it influenced the very idea of time. As Renate Siebert writes in her analysis of the differences of memory of Fascism and Nazism in Italy and Germany, 1968 gave 'our orphan generation' a new perspective on the past which was 'less disorientated and desperate', so that 'we could plan for the future and hope'; its failure was translated by part of that generation into individual choices that perpetuated despair, such as drugs, suicides, and terrorism. Belonging, politically if not always biologically, to that generation, we understand all too clearly the huge effort, in theoretical as well as moral terms, that the collapse of the Communist regimes implies, and the need for reflection on the whole heritage of Marxism.

The second reason is that observers have evidenced the risk of a new cancellation concerning the October Revolution and other historical events. Should the memory of the resistance of the people of Leningrad during the war be cancelled, or obscured, together with the name of that city? To reconstruct churches, and even whole boroughs, as well as to rename streets and towns is not sufficient to revitalize a memory that has been killed. All this is important, but the desire to do away with the past seems to perpetuate the violence of the present on memory; to act as if a large part of the past had never taken place. For the first reason, mentioned above, we feel that the West has a vital interest in giving the past what it deserves, differentiating the various parts of its heritage: for instance, differentiation between values/ideas and their historical implementation on the one hand, as well as between experience of various types of political and economic conditions. It is in this direction that we are working: how oral history can help in indicating differences and similarities in the memories of different totalitarianisms. Therefore we hope to use the concept, not as a cover, but rather as a challenge to understand variations of subjectivity in history.

In one sense, memory seems to have a flattening effect on the concept of totalitarianism, especially in as far as it recalls different experiences in very similar ways. A striking example of this can be found in the recurrent slips by people interviewed in East Germany, who kept confusing names of mass organizations during the Nazi period and during the Stalinist era—for instance, the Gestapo and State Security (see Ch. 4). But comparable mechanisms also appear in the memories of people of different countries: for example, the contradictions between general remarks denying one's participation in Fascism or Nazism and anecdotes implying that the same person actually accepted help from agencies of the regime or consented to certain aspects, such as the cult of the leader, considered as something different from the regime as a whole—Mussolinism differentiated from Fascism; Hitlerism from Nazism; Stalinism from Communism.[21] The direction of study that looks for similarities must be pursued, while more often making use of comparative approaches, still very unusual in oral history.

In another sense, thanks to the researches that have been completed so far, we can point to a number of differences between totalitarianisms as indicated by memory. We shall pick the case of Germany again, thanks to the comparisons that can be made between the project on life experience in the GDR (see Chs. 3 and 4) and an oral history project in the Ruhr.[22] The comparison indicates that the way people remember National Socialism is very similar in the East to the West, but there are some significant differences between the memories of West and East Germans: in the former, the Jews play a prominent role as victims, while in the latter, this role is taken up by the Russian prisoners of war, forced labourers, and political resistance fighters. Such differences can be attributed to a number of reasons, among which are the differences in political discourse between the two Germanies, including a more economistic view in the East, as well as different collective senses of guilt (see Ch. 4).

The length of time seems to have had a different impact on memory, especially if combined with the fact that different generations came under the effect of total domination. That the Italian Fascist regime lasted for twenty years, the Nazi one for more than a decade, and the Stalinist system for seven decades, produced relevant differences for the memories of those who lived under them. Italian militants born before 1920 could, when interviewed by oral historians, retell their life stories as if there was a gap or a parenthesis in them. At the same time, non-political people of the same age could narrate their biographies as if nothing particular had happened during those two decades; yet, their different types of silence were broken by jokes and contradictions, showing that memory has its own ways of recording and transmitting daily life and its relationship to power. But a similar structure of memory cannot be expected in a situation like the Russian one, especially for people of different age-groups, for whom the experience of the dictatorship covered the whole of their lives, there being neither a 'before' nor an 'after' of living under Stalinism.

At the same time we must expect similar mechanisms to appear even with such different experiences, for instance the impact on memory of feelings of guilt and complicity. A widespread attitude of victimization can be found in testimonies of people who lived under totalitarian regimes, expressed by laying the blame on power and on their own helplessness, as if nothing could be done to resist domination. Significantly, this attitude developed only recently in countries like Russia, under the influence of *perestroika*, but especially following the treatment of the matter in books, films, the media, and public discourse in general; even more significantly, this widespread mechanism of memory is reported to be resented by those who suffered more directly, through imprisonment or harassment, from Stalinist repression. A similar resentment is expressed sometimes by Italian militants when confronted with stories by 'ordinary' people about the 'impossibility' of rebelling. But certainly the issue is enormously complicated by the specificity of oppression in the Soviet Union, where it repeatedly happened that groups of people who had been the agents of oppression became its victims in later periods, like members of the party or the KGB, in the passage from the 1920s to the 1930s (see Ch. 6).

In addition to all this, we know of the different effects that the same experiences can have on people of different ages. The attitude of adolescents to totalitarian regimes often seems to have been one of great enthusiasm when they were raised during those regimes, with all the bitterness of discovering the 'lies' behind the appearance in later stages of youth and adulthood. In turn, the same experiences might have had different meanings and relevances, according to the structure and nature of the family and its relationship with the political power: To what extent, for instance, would a Jewish family in the Soviet Union (or a Catholic family in Spain) share a common memory? Or, on the contrary, could the children have experienced opposing loyalties to the regime and to the father, the two authorities that might at certain points be at odds with one another?

It is not by chance that the issue of generations, biological as well as political, and of transmission of memory between them is such a crucial one. We may say that in this field oral history has advanced, and in this book there is evidence of such an advance. Ackelsberg gives a stark example of what happens to the legacy of the revolutionary in a period of political repression; she shows that veteran *mujeres libres*, who had been militants in the organization founded by anarchist women in 1936, and young ones, who took up the name of the organization when they began to mobilize themselves into feminist groups after the death of Franco, proved unable to work together or even talk without degenerating into mutual recriminations and misunderstandings. But she also indicates that the women of *mujeres libres* whom she interviewed had managed to communicate the power of their commitments to their children, who had, in turn, made them their own. No similarly consistent pattern of 'successful transmission' of value could be found among the men, perhaps because women remained the primary child-rearers, even among committed anarchists.

Wierling shows that the generation of East Germans born in the 1920s were affected very differently by the debacle of the 1940s than their parents. While the very elderly tended to collapse personally in the wake of societal collapse, the middle generation appeared to be affected by the period of National Socialism and the Second World War. At the same time, however, the sense of responsibility for their children, and the knowledge that half of their life still lay in front of them motivated them to seek for modes of continuity in their biographies and memories. Thus a direction appears that has not yet been explored in oral history: the relationship between memory, and the future. People's memories of their own past lives, what they remember and what they forget, are shaped by their own expectations for the future, and also by whether they have children or young people for whom they care and who may outlive them. Hopes, fears, and projections converge into shaping memory and its strategies. Oral historians have observed the phenomenon time and again, and are observing it nowadays in the areas that are opening up for research on memory: people who now feel free to narrate their experiences under Stalinism do face the problem of justifying themselves, and at the same time of establishing some sort of continuity in their life stories.

The other extreme in this way of reasoning is the hypothesis that experience at a given time preshaped or restricted future perceptions and spheres of action (see Ch. 3). This raises the question of to what extent one's memory can be linked with collective memory. Niethammer's work shows the connection between perception, political involvement, and memory; it also indicates that the actual perception must have been far more complex before the observer settled on one particular perspective, and established it in the niche of his memory. Therefore, it is not surprising that for people subjected to totalitarian experiences, the content of protest issues was not only repressed as a public issue, but also lost its salience in private memory. And yet, it seems that the moment of public liberation of private feelings is unforgettable. The particular relevance of the overlapping of public and private experience as a landmark of memory had already been pointed out by oral historians at the International Conference held in Amsterdam in 1980: on that occasion it had become apparent that, by contrast to the silences observed in memories of Nazism and Fascism by Anne Marie Troeger and myself respectively, the experience of the Spanish civil war—a time of close cohesion between public and private experience—was absolutely vivid in the testimonies collected by Ronald Fraser for his *Blood of Spain*.[23]

In the debate that followed the conference, the *History Workshop Journal* editorial collective suggested a comparison with the silence about the general strike of 1926, which they had found in the memory of the English working class.[24] They used this example to point out that my explanation for silences in the memory of Fascism could not be generalized, since politics touch individual lives in many different ways. My hypothesis had been that Fascism had accentuated the gap between the political sphere and daily private life, thus creating wounds in the tissue of memory, which could not easily recompose what had been forcefully separated; on the other hand, the testimonies collected by Ronald Fraser sprang from the opposite situation, and resulted from the overlapping of the different spheres of reality in which the civil war had operated its capacity to give meaning to the past, and its link with the present. That debate has not been carried far enough, and there is much space left for further and deeper discussion. What should be stressed at this point, is the importance of the many different types of resistance that oral history has evinced from popular memory of Fascism and anti-Fascism: not only the structured, directly political, clandestine resistance of the militants, but also the daily, cultural, ambivalent opposition—often mixed with forms of acceptance—that nevertheless obliged the totalitarian powers to attend to negotiation and continuous surveillance. On this point, oral history has indicated that contrasts exist between historiography and memory, in the sense that each one of the two tends to forget some crucial aspects, and that reciprocal critique and collaboration are essential to the understanding of history.

In spite of these achievements and the promise of more by research at present under way, the general context is neither a happy nor an easy one. We are faced with very difficult tasks, and sometimes we doubt whether we are up to them. Consider the consequences of two of the articles presented here. Stern has found that Jews who survived the Holocaust emerged from something that he defines as 'borrowed time': they now feel they live among strangers, and the estrangement is reciprocated by many Germans' memories. In fact, the alienation between the two worlds of experience has led to antagonistic memories: the image of the Jews as perceived by many Germans remains highly stereotypical, as part of the continous mental flow typical of the *longue durée* of mentalities; while the self-image of many Jews—as well as their image of the Germans—will never be the same after the unprecedented historical rupture they experienced. In Chapter 5, Khubova writes that she did not find a common collective memory in the Kuban. If that is true, it is a very painful finding. Certainly, before accepting such a conclusion, a historian should go to every extreme in order to disprove it; and indeed, individual—if not collective—memory might have preserved some traces, however pale and vague. The risk is, if such an effort of search is not thoroughly done, that historians cancel for the second time whatever trace of memory might exist, and by this second cancellation definitely confirm the victory of the oppressors. In that case, nothing would be left but mourning. We too would have to grow dumb and silent, and recognize our impotence. This sometimes happens to history; history too can be killed, like memory, and then what else could we do but start from another point and take up the search once again? And yet, there would still be the possibility of a work of *pietas* like that of Liliana Picciotto Fargion, who collected the names of the thousands of Jews deported from Italy by the Nazis.[25]

But we must acknowledge that we actually run the risk of silence—we as Europeans; not merely as oral historians. The 'angel of despair' who speaks the lines by Müller, at the start of this introduction, is the angel of our times: he gives with his hands anaesthetics and narcotics, the drugs of oblivion; his discourse is silence and loud sound, not the word. The full meaning of these lines can only be appreciated when they are sung accompanied by Rihm's music, which breaks and lacerates them to portray the anguish of the contemporary discourse and the agony of communication in it, made more acute by the contrast between huge technological possibilities and increasing forms of racism and hatred among peoples.

Opposing the silence and fury which the angel proclaims, and which are all too evident in wars, whether within or between countries, from the Persian gulf to the violence of youth against themselves and others, what strength do we have? As oral historians, we have frail but precious tools: attention to language and form, to how things are remembered, or forgotten; and not only to the contents of memory, but also to what is not remembered, to silences. We have developed ways of analysing stratifications of orality and memory, and put into practice combinations of content analysis, hints from literary criticism, and insights from psychoanalysis and anthropology. An example of how one can work on memory, going beyond the factual aspects of a story, and yet starting from them, is the article by Kovacs (see Ch. 7). He has noticed that his interviewees do not simply remember the common core of a story but richly build it up, so that at times one account differs widely in content from another, but structurally they are all very similar, even though specific details sometimes plainly contradict one another. This is because the concrete elements of memory are arranged around the same dramatically significant turning-points, and because recollection has two levels: a content-concrete level and a structural-organizational one.

The importance of language is brought to us in the most striking way of Leydesdorff's essay, in which she reminds us of the relationship between interviewee and interviewer, on the basis of her own experience. She interviewed a woman living in Haifa, who had emigrated to Israel after a long period of work with very poor children in Bergen-Belsen: Clara could not talk about the way in which the children she had worked with had been hungry, nor could she give any other information about their desperate lives, other than by using the official language available at that time for charity workers, saying that they were good children and only needed a little help. The story was told in a language that had not been hers at the time: her success in creating a home for those children had been a result of her lack of distance, and her immediate solidarity. Yet, her memory was overshadowed by a feeling of loss: all the children were dead, and it was a hopeless effort to transmit some notion of a devastated culture to an interviewer who was much younger and had never lived in a ghetto.

In the face of such problems, the naïve claim of oral history, in its early decades, to simply give voice to those who had been silenced by history is almost derisory. Fighting silence is not enough; 'silence' is not even an appropriate term for the task to come: what is to be fought is not only silence but distortions or 'false memory'. Part of the memory that is emerging now, because it is no longer impeded, is sometimes just as 'false' or 'wrong' as the one it opposes: suffice it to think of the anti-Semitic traditions in many countries of the Soviet Union, where the rehabilitation of anti-Stalinists sometimes risks rehabilitating anti-Semitic or even Fascist and Nazi people. The problem is that even 'tradition' is not enough. As Adorno pointed out many years ago, the very idea of tradition has been questioned and put out of date by the developments of capitalism, in which traditions become aesthetic objects for consumption and mass tourism; charming but dead traditions ('la beauté du mort'), as indicated by Michel De Certeau. What is even more deadly is to try and revive them rather than seeing the depth of their destruction and the frailty of their revival, which does not take into full account the distance between their original context and our present one.

The destiny of a traditional memory takes us back to our beginning and to the story with which we started. Now that we are near the end of this introduction, I should like to pose the initial enigma once again, by means of another story. In the *Schachnovelle* (Chess Novel), which Stefan Zweig wrote a few months before committing suicide together with his second wife, in Brazil in 1942, the player who has based his survival from Nazism on his capacity to use memory is defeated by a human automaton, somebody who understands nothing but the game of chess. The former is an Austrian lawyer, who had been imprisoned by the Nazis in a hotel room for four months because he was trying to hide some financial secrets belonging to the Imperial House. They were four months of no human contact, undergoing torture and interrogations, and nothing to do or read. But he had succeeded in stealing a book that happened to contain a series of great chess games; he had learned and memorized them, thus saving his mind and memory, while fooling the Nazis by his resistance. However, he now succumbs to the mechanical player: he—the cultured, refined, humorous European—is defeated by a man of very poor origins—the son of a Danube boatman, who had been deprived of any culture, ethics, sense of humour, and self-irony. Doctor B. is a real *dilettante*, in the sense that he draws *diletto* (pleasure) from playing; Czentovic is the professional, interested only in winning and making money out of his victories. The former loses because his memory goes berserk under the stress of competition, and he mixes up the game he is now playing with another game that he used to play a long time ago.

The story has its strongest impact on readers by making them feel violently threatened, as if it had a prophetic value and was announcing the deadly risk of people losing everything they value. It is understood that, culturally, readers belong to the European middle classes and cherish their traditions; these traditions, and memories, can no longer compete when the rules of the game are those of the contemporary world. They were enough to stand up to Nazism, but are no longer apt to confront a progeny of Nazism, that is the new illiteracy of robots. It is perhaps not by chance that the fatal game is played on a ship sailing to the USA and leaving Zweig's *geistige Heimat Europa*, his spiritual homeland. No wonder the memory of Europe appears unable to save its values: it shows here some of its strongest limitations, in terms of class values and hierarchy of human beings. However, we can perhaps read the *Schachnovelle* from a new perspective after the collapse of Communism: the outcome of the game between the master and the slave could not save Europe; in other words, the Eastern proletariat would not foster more hopes than the Western *bourgeoisie*—on the contrary, it would impose even worse forms of oppression. To the cold eye of the future the brutal Slav and the refined Austrian rise and fall together, since the one who wins today, and had been a victim for so long, has learned his lesson: he is victorious because he has discovered the market, and now knows its ruthless laws better than his opponent, who does not realize that everything, even leisure and games, has been subjected to those laws. In this way the master of yesterday gains in our eyes a moment of truth, in as far as he retains a touch of humanity that the present is destroying. But our memory cannot be reconciled with the image chosen to symbolize the robot—the Slav, the poor, the peasant—who, like millions of other migrants to the New World, should have had another chance, rather than displaying his shrewdness on the capitalist market.

A possible epilogue could be composed mainly of hopes for the future, while trying to look at the meaning of the *Schachnovelle* from a different perspective. It is not merely by chance that the connection between memory and otherness has been emphasized by those who have tried to reformulate the heritage of Europe in the present world, and to bring democracy 'up to date'.[26] The hope is that the memory of Europe can be reformulated in such a way as to be comprehensible, even if not acceptable, to people of all social groups in all parts of the world. Today, it is the multiplicity and diversity of memories, as well as of languages, that constitute the only privilege of Europe. It is the closeness of so many different forms of expression in such a narrow space and with equal dignity that memory will have to reflect and transmit. Memory—or better: memories—can help us to find ways, in the era of equality and cosmopolitanism, 'to participate in the other, to share his/her being other'.[27] For us, the task is to participate in different memories, to share their differences not in any way in an attempt to demonstrate their universality but rather to insist on the diversity and plurality of memory. Engaging in this task, inventing ways of facing up to this challenge, will be the contribution that oral historians along with all those concerned with the processes of remembering can make, in order to detach human memory from all forms of totalitarianism, in politics as well as in culture, and to help it play its part in the forming of a democratic consciousness.

Notes

1. The lines by Heiner Müller are part of *Frau/Stimme*, for soprano, and orchestra with soprano, composed by Wolfgang Rihm in 1989.
2. *Memoria* was a joint production by the Nordisk Theater Laboratorium of Holstebro, Denmark, and the Teatro Tascabile of Bergamo, Italy. I saw the version performed in Turin in Jan. 1991: it was directed by Eugenio Barba of the Odin Teatret, with Else Marie Laukvik and Frans Winther as actors; Yiddish folk-songs and music were by Frans Winther.
3. Jean Améry is the pseudonym of Hans Mayer, born 1912, a Jewish writer in the German language. He was deported to Auschwitz as a member of the Belgian Resistance and committed suicide in 1978.
4. Paolo Jedlowski, 'Il testimone e l'eroe: La socialità della memoria', in id. and Margherita Rampazi (eds.), *Il senso del passato: Per una sociologia della memoria*, Milan, 1991, 15.
5. Patrice Meyer-Bisch in *Forces et faiblesses des totalitarismes*, Actes du IV Colloque sur les totalitarismes, Fribourg, 1987, 108 ff.
6. Norberto Bobbio, *Il futuro della democrazia: Una difesa delle regole del gioco*, Turin, 1984, 4–6.
7. Id., *L'età dei diritti*, Turin, 1990.

8. Edgar Morin Ernest A. Menze (ed.), *Totalitarianism Reconsidered*, Port Washington, 1991, 22.
9. Peter Weiss, *Die Ästhetik des Widerstands: Roman*, Frankfurt am Main, 1978; trans. *Esthétique de la resistance*, Paris: 174.
10. Paul la Hausse, 'Oral history and South African historians', *Radical History Review*, 46/7 (1990).
11. *Oral History Association of Australia Journal*, 12 (1990).
12. Jean Widmer in *Forces et faiblesses*.
13. Stephen J. Lee, *The European Dictatorships 1918–1945*, London and New York, 1987, xi.
14. Menze, 'Obedience and human automony', in id., *Totalitarianism*, 234–5.
15. Maria Ferretti, 'La memoria mutilata. Storia e memoria nell'Unione sovietica degli anni ottanta: La società sovietica alla ricerca di un'identità', unpublished manuscript.
16. Mikhail Rojanski, 'Tsaritsine, Stalingrad ou Volvograd? L'éternal débat sur les toponymes', in Alain Brossat, Sonia Come, Jean-Yves Potel, and Jean-Charles Szurek (eds.), *A l'Est la mémoire retrouvée*, Paris, 1990 (reviewed here).
17. Pierre Nora (ed.), *Les lieux de la mémoire*, Paris, 1987.
18. Jacques Le Goff, 'Préface' to *A l'Est la mémoire retrouvée*, 9.
19. This term, from Ernst Bloch, is taken up by Brossat *et al.*, 'Introduction' to *A l'Est la mémoire retrouvée*, 21 ff.
20. Morin in Menze, *Totalitarianism*.
21. Luisa Passerini, *Mussolini immaginario: Storia di una biografia 1915–1939*, Rome and Bari, 1991.
22. Lutz Niethammer (ed.), *'Die Jahre weiss man nicht, wo man die heute hinsetzen soll'. Faschismus-Erfahrungen im Ruhrgebiet: Lebensgeschichte und Sozialkultur im Ruhrgebiet*, Berlin and Bonn, 1983; id., (ed.), *'Hinterher, weiss man, dass es richtig war, dass es schief gegangen ist'. Nachkriegserfahrungen im Ruhrgebiet*, Berlin and Bonn, 1983; id., and Alexander von Plato (eds.), *'Wir kriegen jetzt andere Zeiten'. Auf der Suche nach der Erfahrung des Volkes in nachfaschistichen Ländern*, Berlin and Bonn, 1984; Niethammer, von Plato and Dorothee Wierling (eds.), *Die volkseigene Erfahrung: Eine Archäologie des Lebens in der Industrieprovinz der DDR*, Berlin, 1991.
23. Ronald Fraser, *Blood of Spain: The Experience of Civil War, 1936–9*, London, 1989.
24. Editorial: 'Oral history', *History Workshop*, 8 (1979).
25. Liliana Picciotto Fargion, *Il libro della memoria: Gli ebrei deportati dall'Italia (1943–1945)*, Milan, 1991.
26. Jacques Derrida, *L'Autre Cap. La Démocratie ajournée*, Paris, 1991; trans. Oggi l'Europa, Milan, 1991, 25.
27. Hans Georg Gadamer, *Das Erbe Europas*, Frankfurt-am-Main, 1989; trans. L'eredità dell'Europa, Turin, 1991, 46.

2

Antagonistic Memories

The Post-War Survival and Alienation of Jews and Germans

FRANK STERN

> . . . back to my territory. We went to the town where I was born. I saw my house where soldiers now were sleeping. Some Germans from the town recognized me, even in uniform, 'Willy,' they called, but I wouldn't answer. In the distance was the cemetery where my mother was buried.[1]
>
> (Julie Heifetz, *Oral History and the Holocaust*)

During Passover 1945, American troops captured Frankfurt and liberated the 'handful of ghostly remaining members of the city's Jewish community'. Meyer Levin, an officer in the American army, recalls that day:

> I went into Frankfurt and found the remaining Jews in a few ghetto houses. There were 106 people out of a former population of 40 thousand. It was strange how the Jewish GIs had already gravitated toward them. But according to military regulations, these Jews were German civilians, and fraternization with them was forbidden. So the GIs had left packages of matzoth for them on the doorsteps of the ghetto houses. The Jewish soldiers watched from across the street as the last Jews of Frankfurt slipped out, still fearful, and picked up the Passover food.[2]

In the vicinity of Berlin, a Jewish nurse survives the final stages of the Third Reich and the persecution of the Jews. Her familiarity with the

This paper, presented at the Seventh International Oral History Conference, Essen, 1990, is based on published and unpublished interviews, memories, and reports collected at the Hebrew University of Jerusalem and Tel Aviv University by the project 'Lebensgeschichte und Sozialkultur im Ruhrgebiet 1930–1960' (Open University, Hagen), and by the author. See Frank Stern, The Whitewashing of the Yellow Badge: Antisemitism and Philosemitism in Post-War Germany 1945–1952, Ph.D. thesis (Tel Aviv, 1988).
Translator: Bill Temple.

German milieu proves useful to her. She manages to get false papers, French forced labourers assist her out of a sense of solidarity, and not all of her former acquaintances turn their backs on her. In addition, she gains some help from former political friends and associates connected with the Communist resistance.

After liberation, she returns to Berlin and starts looking for her 12-year-old daughter, who has been living semi-legally with her anti-Semitic mother-in-law, without any love or affection. In the beginning, the girl rejects her mother, and wants to stay with her grandmother. She mistrusts her mother, who has emerged from the horror of the camps and bears the stigma of a Jew. If only her mother had not abandoned her and given her to her grandmother, who hated her and beat her repeatedly. The girl is confused, troubled, helpless—and yet defiant. Her mother approaches her cautiously, waiting until her daughter gains confidence.

> It wouldn't have been all that bad. But when the Soviets came, the old woman told her that her mother wasn't worth a damn. She hung a sign on her: 'I'm a Jew. My mother is Jewish, she was gassed in a concentration camp!' That's how she sent her to the Soviets to beg for something to eat . . . So the old lady didn't want to hand her over. I tried to bargain with her . . . Everyone who came out of the camps back then got a lump sum of 400 marks to make a new start. Well, I gave her my 400, along with my ration cards. And she gave me my child back in return. It's easy to tell about it, but it was a difficult experience. So I got my child back. Whatever energy I had I invested in the child . . . She had never forgotten that as a very small child once, wearing a Star of David, she had been beaten bloody. And after a short time she said she wouldn't have it, she wasn't going to go to school with German kids.

The girl is accepted into a Jewish youth home in northern Germany in 1946, and then emigrates to Palestine in 1947. Her mother remains in Germany, living until her death in the GDR. Besides all the problems of a family shattered by racial persecution and riven by long-standing tensions, this quotation indicates the manifold personal impact of the historical situation on the encounter of Jews with Germans in 1945.

There is a report by Martin Riesenburger, later a rabbi in Berlin, that a number of children came to him immediately after 8 May 1945, when he had advertised on walls and on the radio for Jewish children to contact him so that he could organize the first classes in Jewish religion in the city after liberation. Already in May, the first Jewish wedding was celebrated, and it was not long before the first Jewish child was born in post-war Berlin. But does this entail continuity of

Jewish life? Rather, it points to the necessity of thoroughly investigating the Jewish post-war experience in Germany and the evolving Jewish–German relationship.

American military offices dealing with the problems of Jewish survivors in Berlin came to the following conclusion in September 1945:

> Of the Jewish population in Berlin in 1933 ... some 6,000 Jews are left in Berlin as of this date, and made up of the following categories: 1,155 released from concentration camps; 1,050 concealed in Berlin; 2,000 who were married to non-Jews and who therefore received some degree of preferential treatment although they had to wear the 'Jewish Star'; 1,600 married to non-Jews, whose children were also raised as 'Aryans' and who did not have to wear the 'Jewish Star'. The total includes about 100 children, 14 years of age or younger. ...[3]

It would be a challenge for future oral history research to interview those one hundred who were then children. However, the emphasis of oral historical research to date has been more on German Jews who had the chance to leave Germany. Indeed, it is a revealing phenomenon that these survivors would appear to be more in keeping with the emotional needs—and scientific priorities—of young German scholars today than that group of potential respondents who survived in Germany, 'inside the vicious heart'.[4]

In what follows, the reports and memories of Jewish survivors in Germany will be set against the Germans' perception of the Jewish remnant in the period of transition from the Third Reich to the German post-war era and the early years of the Federal Republic.

In the winter and spring of 1944/5, a remnant of German Jewry and some 50,000 Jewish forced labourers had survived Nazi persecution and genocide within the very heart of the Third Reich. Approximately 15,000 German Jews had survived in the 'Altreich', that is the German state territory within its 1937 borders. This was a mere fraction in comparison with the more than 500,000 Jewish citizens of the country in January 1933. The history of that remnant of German Jewry, that miniscule minority of survivors, still needs to be researched in depth, and has generally been ignored by German historiography. Few autobiographies or published interviews provide information about this chapter in German post-war history.

On the other hand, there are quite a few historians and social scientists who follow the traditional philo-Semitic guiding principle: namely, that one should focus in particular on prominent Jewish artists, writers, and scholars in interviews or historical descriptions, as if it were still a matter, now as before, of lauding the 'Jewish contribution to German culture and science'—as if that would prove anything other than the presence of a bad conscience.

While the history of German everyday life and popular oral history have been dealt with in a multitude of projects and publications, the Jewish dimension of post-war German history generally continues to appear in a stereotypical form. On the one hand, there is the figure of the returnee, who is duly integrated into the conservative consensus of the Adenauer era—presumably forgiving his or her former persecutors, and guided by a 'spirit of reconciliation'. On the other, there are the Jewish displaced persons from Eastern Europe, who become catalysts of anti-Semitic continuities in the private and semi-public sphere—persons to whom Germany must appear as a country of transit, and who are indeed publicly confronted with just this image of the country. In descriptions of regional history, one finds mention of Jews who 'disappeared' before or after the November 1938 pogroms, and then after 1945, if at all, the re-establishment of synagogue congregations. Given this background, Jewish experiences within the German *Nachkrieg* can be conveniently dealt with as a history of representative Jewish institutions, and can thus be reduced to institutional aspects of political history. The focus then need only be on the confrontation with a supposedly institutionalized minority. If one leafs through the mass of publications issued on the occasion of the fiftieth anniversary of 9 November 1938, the fortieth anniversary on 8 May 1945, and the fortieth anniversary of the establishment of the two German post-war (and obviously transitional) states, one rarely finds that surviving Jews—with their individual and collective experience—are granted a proper place in these representations of collective memory.

Historical consciousness appears here to be segmented, and the separation, the divide wrought by the Third Reich between German and Jewish history extends far beyond the first post-war decade. The shift from an open discussion of collective national guilt in the spring and summer of 1945 to, on the one hand, the proclamation of the first President of the Federal Republic of a 'collective shame'—a substitution which was an evasion rather than a true 'mastering of the past'—and on the other hand, the rejection of any guilt or responsibility for the previous Nazi decades by the new Communist GDR—as a prelude to its own collective amnesia—was on each side a notably quick step in the development of post-war German national consciousness.

This transformation can be contrasted with the memory of German Jews who lived and survived in the Altreich. These 15,000 Jews outlived the Shoah (catastrophe), and required great mental and physical strength to survive what they were confronted with in post-war Germany. They resisted, and in 1945 and thereafter they had to accept the stark reality that there could be no simple, straight road leading back to the situation as it was before 1933—that one could not simply go on living 'as though nothing had happened'. In a process that was social, cultural, economic, and political, the majority of German society had not only revoked the civil rights of the large Jewish minority in the country, it had also abrogated their very right to exist. The massive and minutely-organized mass murder, involving active participation or silent complicity on the part of millions of Germans from all social strata, had resulted in the physical annihilation of German Jewry. There was—and is—no analogy for this historical fact; for the few survivors, there was no point of reference or orientation to help make sense of their individual and collective historical experience. They left the camps, their hiding-places, and lived semi-legally or underground, and returned—yet unlike the great mass of Germans, not to their homeland. Rather, they came back to a 'German place that was foreign'—a 'deutsche Fremde', as it was dubbed by Hans Mayer, one of those returning from exile.

Encounters between Germans and Jews already took place before the liberation, and were intensified afterwards under the Allied occupation of Germany. The memories of Jewish survivors and Germans reflect the character of mutual relationships which developed against the background of both the former official genocidal anti-Semitism and the all-embracing post-war crisis of German society and culture. The analysis of continuities and discontinuities in the perception of German–Jewish relations on both sides leads to my main thesis: the emergence and deepening of antagonistic perceptions and attitudes in everyday life which are—even today—at the core of German–Jewish relations.[5]

Rudolf Schottlaender, philosopher and philologist, survived in a 'privileged' marriage in Berlin. At the end of 1942, he took on 'work as an unskilled labourer at the lowest level' in a powder factory. About his experience toward the end of the Nazi regime, he recalls:

> In the beginning, there used to be half a dozen Jewish fellow-workers who would have lunch together with me in the factory. After they were deported, I was left there alone. Me, a person who loves to talk when eating a meal, for

years I ate all alone out of a lunch-box brought from home! Because I wasn't supposed to have any contact with people . . . A Social-Democratic worker who was convinced the war was lost tried to have conversations with me. For a while that went OK, until our extra-tall supervisor got wind of it. He forbade both of us to talk—me, in a threatening tone. I replied to him: 'If I'm not allowed to talk to anyone any more, then I might as well be dead.' From then on, he left me in peace. My reply had apparently impressed him. Yet I didn't provoke him either by being very open about breaking the prohibition. Once I got scared. A worker above me in rank, one of those old union types, always very objective, never anti-Jewish, was walking next to me, and we could see the flames from an air raid on Berlin. Suddenly he blurted out: 'Those are your friends up there!' All I said was: 'I didn't invite them.' Yet I was deeply affected by this, since the alienation from all the others, which I was aware of as a daily thing, came this time very unexpectedly, and had been directed toward me personally.[6]

In 1934, the Comité des Délégations Juives in Paris had described the situation of Jews in Germany as a 'revocation of the emancipation of the Jews'.[7] After being stripped of their rights, they had been discriminated, humiliated, brutally persecuted, deported, and then murdered. Corresponding to a tendency toward accommodation with collective injustice on the German side, Jewish reaction was a mixture of diverse components: despair, hopelessness, escape into suicide, the will to survive, and a *generally unbridgeable alienation*. The small number of Jews still remaining in Germany were in a situation where they had to rely on themselves. Over a period of twelve years, they had had nothing but the fact of their 'Jewishness', however that was understood or rejected, and the humanity shown by a small number of Germans. The 'alienation from all the others' described by Schottlaender is the decisive factor in relations which had once been so emphatically extolled as the 'German–Jewish symbiosis'. This holds true for the Third Reich, as well as the transition to post-war Germany, and the Federal Republic.[8] It applies both to Jews forced to wear the yellow star and to those placed in a different category by the racist madness. In other memories, this alienation is termed as 'an invisible wall dividing us, one which both sides were dimly aware of'. And the other side was perceived as 'the normal people'.[9] The Germans had decided that their Jewish neighbours were alien, strangers, and now it was the Germans themselves who became total strangers in the perception of the small number of surviving Jews inside Germany.

The final collapse of the Nazi empire approached, and many Jews heard more and more information about the extermination camps from Germans, soldiers on leave, and others. This shaped their feelings about their German environment, and they questioned their future identity even before liberation. A young Jewish woman, who survived by living in a monastery near Freiburg, expresses the feelings of conflict and guilt that could be associated with survival, with the necessity to survive:

Back in those days, the last of people like me who were still in Germany, Jewish spouses in mixed marriages, were arrested and sent east: the place where my grandmother went to her death, where my friend was turned into a pile of ashes that no one saved. That was where my path was actually heading; that was my place too. Helpful hands had taken hold of me, hidden me, so that I lived a peaceful life here, in hiding, protected by days that were unaware. Yet this life didn't belong to me. I was living on borrowed time. I could use that time, but it wasn't mine. Life was generous. It allotted the stranger his hours just like any other living thing. It bestowed on him its gifts: to wake up in the morning, four walls and a roof, one's daily bread. It let the evening come. It was generous, this strange life. And it didn't even seem to notice how it was bestowing its gifts on someone who actually had no claim to them. But I had to be honest and sincere. My soul withdrew from that alien, borrowed peace and wandered eastward. To that place of its allotted fate: toward calamity and ruin.[10]

In this connection, it should not be forgotten that German Jews still living in the Reich at this time frequently made use of direct personal contacts—along with their familiarity with the language, their education, and their knowledge of the social and political scene—in order to escape the threat of imminent deportation.

A young Jewish woman from Berlin, who had gone underground shortly before the 1943 'factory action' and then spent months hiding in telephone booths, waiting-rooms, and public toilets, tells about her experiences during the final months and weeks:

Now the bombing raids became heavier and heavier . . . and terrible as that was, the general commotion associated with all this somehow had a favourable effect when it came to those living in hiding, illegally. It was easier to 'slip away' when necessary, one was no longer so conspicuous, everybody had enough of his own worries . . . I scraped by, taking care of children during the day, and working as a cleaning-lady and casual labourer. There was hardly any arrangement for sleeping that I hadn't tried out—on verandas, in backyards, in attics, under the big iron in the laundry and in stairways. . . . Finally though, I was unable to escape my fate . . . I was put in the camp on Schulstrasse, and remained there until liberation, for ten months. There too, I was fortunate. I was given work in the household of the head of the communications office. Up until 22 April 1945, when things got too hot for our Gestapo guards, and now they were the ones who started 'scurrying', just as I had 'scurried' to stay alive—though in my case I hadn't committed any crime. It was only because I'd been born Jewish![11]

Siegmund Weltlinger, who survived together with his wife in hiding and was one of the founders of the Jewish Community and of the Society for Christian–Jewish Co-operation in Berlin, recounts similar experiences:

We never left the small room, not even during air raids; because there were a lot of zealous Nazis living in that block of flats, and we knew we mustn't endanger our benefactors. Once there was a direct hit on the house next door, and part of the wall of the room collapsed from the air pressure. We were not injured. The bombs didn't frighten us as much as the fear of being discovered or badly injured. Because what should we do if that happened?[12]

The experiences during night-time air raids are some of the most penetrating and central factors of the antagonistic memory. Of all the memories and tales of the ordinary German population, the air raids, destruction, and deaths represent a wartime hell that dominates everything else. The bombing raids were *the* great catastrophe, and in later recollections and stories constitute the main element in the description of German suffering, even more so than guilt and expiation, responsibility, and the crimes of the Third Reich. These recollections reduce individual participation in the National Socialist community and polity to a collectively shared suffering—without asking about its causes. During those nights, under aerial bombardment, even before the end of the war, the equality and moral identity of post-war Germans had already been fashioned. That notion was manifest even many years later in the countless calculations that equated the victims of such air raids with those murdered in the extermination camps.

Yet for the small number of Jews, despite all their anxiety and fear, those same air raids constituted above all else a ray of hope. The ruins and the partial chaos weakened the bureaucratic grip of German officialdom. It was thus possible to get hold of documents that would allow one a false identity and semi-legal possibilities for residence. Here alone one can clearly see just how dubious any description of everyday life or oral history of this period is if it excludes the Jewish side of experiences in this transitional phase from war to post-war.

The liberation was a military, social, and political process which took place at a very uneven pace on the various different fronts. Though it principally had a collective character for the survivors in the camps, for many Jews who had been in hiding in the Reich itself liberation was by no means the end of their isolation. The young Jewish woman, quoted above, who had been in hiding in a monastery describes her path toward freedom, emerging from the walls of the monastery one day after the arrival of the French troops.

The next day I left the monastery and wandered back to town . . . It was freedom! Then someone came toward me. A man from town, with a rucksack on his back . . . The first human face I had seen in my newly gained freedom: the enemy. I saw many, many more faces standing behind him: men, women, children, a vast throng. They walked through their streets, rode on their streetcars, led their lives behind their windows—just as always. Then fear descended upon me from the open heavens and grabbed hold of me as I stood before this face. He was there now: the enemy, in the flesh. From the world I had built for myself, I returned to that of the others: the world that now once again was to become our common world. Now we were to walk and ride together once again. We would visit each other's homes, I would buy at their shops, and we would meet at the theatre. As if nothing had happened. As if they hadn't stood up from their office desks and stepped out of their living-rooms in order to sweep the pavements clean of us. Now they returned to their offices and houses, and their pavements were open again . . . There was no return. I knew at that moment that things could never work out again. The hatred, the rifle butts, and gas had separated our lives and theirs. Our eyes would never be able to meet again. Even if goodwill should arise from the depths of their hearts and move forward to meet our longing: to be one of them—the last word had been spoken about them, about us. Things would never be all right ever again.[13]

On the morning of the liberation, the Jewish woman from Ahlen, mentioned above, acted as a go-between for the American soldiers and the family of the farmer who had hidden her. Her husband left his hiding-place, and the family returned to the town of Ahlen, from which they had been expelled seven years earlier.

Our joy at liberation was darkened because none of our relatives returned. We stayed alone. Soon our dark premonitions turned to bitter certainty: we had lost thirty-seven close relatives. Other Jews, likewise, did not come back. The Jewish community in Ahlen had ceased to exist. We no longer had the strength to hate and take revenge. We weren't the only ones to feel like this, there were others too. Wherever Jews resurfaced, it was always the same: they were too hardened, too weak to feel pure and unmixed joy as a result of their regained freedom. There was not even the strength for hatred, for feelings of vengeance. After twelve years of being oppressed and outlawed, we were no longer capable of any heroic feelings . . . Vengeance was not taken against any of our tormenters, even if we knew his deeds full well. Since then, I've wondered a thousand times whether that was really and truly the end. The end of what?[14]

The transition from the Third Reich to the German *Nachkrieg* is certainly not a 'zero hour', but it is the deepest rupture, the most profound *historical break* in the whole of German history concerning the relationship between Germans and Jews.

Nothing expresses the fundamentally altered situation in the Jewish–German relationship better than the reactions, recollections, and reflections of the liberated. The thoughts quoted above are just as symptomatic of the situation as is the widespread and profound feeling of alienation mentioned by many other survivors in Germany in May 1945. What they experienced was not the invasion of a victorious enemy but liberation, not the common experience of defeat and occupation but individual survival and total social isolation, not collective solidarity by the Germans but distancing helplessness. The exceptional circumstances under which these Jews survived was their common denominator. They had lived on borrowed time. They knew this, and they took that burden with them into the German post-war period. By dint of their experience, the Shoah, the Germans had become strangers to them. The surviving German Jews gave witness to the historical rupture in the history of German Jewry. After 1945, this part of world Jewry ceased to exist, although there remained *Jews in Germany*.

What characterizes the German memory of this transition in 1945? When reading the vast number of interviews with German inhabitants of the Ruhr area collected by the project on 'Life Histories and Social Structure in the Ruhr, 1930 to 1960',[15] it would appear that relationships with Jews and personal experiences of the interviewee in connection with Jewish persecution tend to be mentioned in a casual fashion, almost parenthetically. If members of the older generation bring up such topics, it is usually because they had direct contact with Jews. It is a different situation with the younger generation, whose representatives speak about their experiences at school in the Third Reich. One can sense here the anti-Semitic pressures of Nazi domestic policy. The exclusion and persecution of German Jews becomes clear through the manner in which interviewees perceive the past in recollection.

When one reads recollections and stories from the age-group of those who were at school during the Third Reich,[16] who were members of the Hitler Youth (HJ) or the League of German Girls (BDM), or who worked as apprentices, or perhaps were even enlisted into the army, then one can note five recurrent thematic areas in such conversations. First, the perception of, or contacts with, Jewish schoolmates or children in the neighbourhood. Such contacts and perceptions took place at an ever greater distance with increasing activity in the Hitler Youth and in the BDM, as well as after graduation from school. Second, the recollection of Jews in connection with the events of the pogrom in November 1938. Third, the recollection of the disappearance of Jews, which even in retrospect is reported in an exceptionally naïve manner. The fourth is the lack of any perception, apart from a few exceptions, of the daily terror perpetrated against Jews prior to the November pogrom and the implementation of 'Aryanization' throughout the Reich. Last, in some cases, the reappearance of Jews in the post-war period—this is usually connected with an intermingling of economic anti-Jewish bias and what can be termed a new 'philo-Semitic *habitus*' which depicted everything Jewish in a usually positive vein.[17]

A woman from Essen, born in 1929, recalls a Jewish fellow-pupil:

> I myself had a schoolmate, she was half-Jewish . . . They came one night and took her father away, and three or four days later she was notified that he was dead. That was better for this girl, because at school she . . . she couldn't get ahead, when she would get a 'B', then the other girls would shout at her: 'Why don't you go to Palestine!' I met her again later on, and she told me that she couldn't stand to look at that other girl . . . Her hands were always moist with sweat because she was frightened.

Apart from the chilling formulation, 'That was better for this girl', what is striking here is that the behaviour of other people is described, but the respondent's own reactions—if in fact there were any—are not indicated. The later encounter apparently took place after the end of the Third Reich. Though she was constantly afraid, the girl had 'attended school quite normally'. To have known that the father of her classmate had been murdered by the Nazis is apparently not in contradiction with the stereotypical post-war profession of ignorance: 'people didn't know what was happening to the Jews'—or with the knowledge they did have about Dachau.

Few of the interviews indicate any behaviour of one's own that might have reflected a sense of solidarity. Nonetheless, there are recollections of dismay, a general human sense of shared sympathy, or even rage at the anti-Jewish measures, which were visible as a form of harassment in everyday life. Occasionally 'the family' helped, or simply 'people'. Generally though, the perspective in such instances is that it wasn't possible in any case to do very much—that one was subject to the circumstances at the time, and the prevailing policies. If other interviews sometimes reconstruct the subject's reaction in recalling some event, then the choice of words oscillates between 'Jewish girl', 'Jewish classmate', 'Jewess', or 'Jew'. With very few exceptions, there is no mention of first or family names. The recollection of the feeling that Jewish girls 'weren't very nice' is often adjusted in deference toward the interviewer to what is recognized as socially correct: she was not rejected 'as a Jew'.

The description of Jews is generally colourless and abstract in its depiction; it moves from a sketch of episodes to a highly diffuse specification of time by means of formulations such as: 'but then one day they just disappeared'.

> As I said, you simply didn't have any further thoughts about the matter. I don't know if that's any different today with 18-year-olds . . . About what was happening around you, you know. Though today there's so much talk about this. I lived in the south in a neighbourhood where . . . a lot of Jews were living . . . And I didn't give it a second thought. Then people said: 'The Levis have left, they've gone to the US.' Or something like that. So I thought: 'OK, they're in America.' As a young person back then, you didn't give such things any thought.

In contrast with other interviewees, whose age perhaps did act as a genuine barrier to more detailed and probing questioning, the 'thoughtlessness' of the interviewee here is probably less acceptable. She was not only a high-ranking functionary in the BDM, but rejected Jews even later on, as other remarks of hers in the interview indicate.

A car mechanic born in 1926, who later became an engineer and was a businessman at the time he was interviewed, describes in detail experiences which, in the form of their presentation, are apparently already a kind of communicative narrative; these experiences seem to be relatively polished in the way they are described, though one can sense certain stereotypes lurking beneath the surface.

But when you'd ask, well, then you'd get these vague answers. Though, as I said, I myself don't know anyone personally who suffered from persecution, who was locked up. After all, we were a people of 80 million. So if it's right what they say, that 20 million were liquidated by National Socialism, that would mean one in every four. I don't know anyone. So if that kid R. ended up in some camp, well I don't know. He just disappeared one day, I don't know any more than that.

A recurrent pattern in the data is that one's understanding of what occurred to the Jews and one's own life history flow on in unbroken continuity from the 1930s into the 1940s and 1950s. Here too, there is no 'zero hour'. There are significant references, which occur in many interviews, to the wealthy Jew, the classmate who disappeared, or the lack of knowledge about the fate and whereabouts of the Jews, and sometimes a hint of questioning about the mass murder. The respondents seldom report questions on why the Jews had disappeared and where they had gone. If questions like that were nevertheless raised, the degree of which even children apparently noticed individual incidents of persecution or brutality in their neighbourhood, on the way to school, at school, or in conversations within the family, is quite astonishing.

Two aspects may play a role here. First, it may be that general experiences at school and on the street induced interviewees to fear reactions to being questioned, engendering a sense of caution: so that one only spoke about such matters in the family circle or among close friends. Second, it is possible that adults made the topic a taboo subject to children, as well as for themselves outside an immediate intimate private circle of family and friends. The official, public opinion influenced the atmosphere at school. Expressions of private opinion were not supposed to lead to any social risk.

This recounting of recollected perception and its later formulation has some importance because of the fact that it occurs so frequently. First of all, the nearly identical phraseology among persons differing in age and origin is striking. What is being related here is not later, retrospective rage at Nazi racism; rather, these are stylized event-narratives which are characterized by *emotional distance*. Sequences are sketched in which 'the' Jews appear, to a significant degree as a social phenomenon bound up with Kristallnacht. And then they 'disappear', they 'vanish' once again from these sequences; as if in a motion-picture film, they are given the social reality of a scene, of an appearance on the stage or before the camera. This may indeed have been in keeping with the actual facts of social reality—perhaps the individual did not know any Jews personally. The relationship with Jews in such a case, quite independently to Nazi propaganda and anti-Jewish measures, may naturally have remained relatively abstract. However, the presence or absence of personal contact do not provide a sufficient explanation for social behaviour in a given instance.

The formulation about Jews 'disappearing from sight, vanishing' is also revealing in respect of the perception of politics and political events at the time—the perception of interference in 'normal' everyday life coming from political circles. Since it was apparently a question of 'grand politics' on a national scale, when it came to racial policies, deportations, and annihilation, the question of one's own responsibility in all this is not raised. One felt helpless, subordinate. As someone states in one of the interviews: 'politics weren't discussed—they were a taboo subject'. One had nothing to say—the others (high up in government) had said it all. One of the main focuses of Allied re-education activities after 1945 has to be seen in this context: the interrelationship between democratic development and the endeavour to combat anti-Semitism among the German population.

Recollections about Jews can contain an expression of disapproval, lack of understanding, sense of helplessness, or a certain kind of limited justification by omitting any evaluative commentary whatsoever. At the same time, the moral weight of the formulation 'disappearance' is disturbing from a present-day perspective, considering how frequently it recurs. The question arises whether in fact they knew more about the nature of this 'disappearance'. But this formulation can also be a reflection of the ideological indoctrination the population had undergone: that the Jew was the stranger, the 'alien element'—individuals who had to vanish from Germany. In other words: if the question is posed explicitly as to the interviewee's image of the Jew—which was, and still is, present among a portion of those interviewed—what is involved here is contact with Jews and Jewish persecution as a passing phenomenon, a transitory reality in the context of one's own personal biography.

In addition, a salient question presses itself upon the researcher: In what way do Jews and Jewish topics crop up again later on? It is conceivable that the *abstractness and vagueness*, the non-questioning of their disappearance constitute conditioning factors for a disposition which shapes the reactions to their reappearance in altered social processes.

It should not be forgotten that the recollections referred to here involve that generation of Germans who were between the ages of twenty and forty in the 1940s and 1950s. At the time of the interview, values and attitudes emerged which were partially new. Yet they simply were asserted; they were anticipated in advance as the opinion which the interviewer held or expected to hear. There is a gap between the emotional—or retrospectively reconstructed—experience with Jews on the one hand, and elements of a rational working-through and processing on the other. In the post-war period this gap is bridged by the new socially-set norm of what one *should say* when speaking about Jews—to comment and speak in positive terms only. The enemy who had to be hated became, in the new public context, the enemy who had to be loved. This influenced not only the use of language in speaking about Jews but also the individual reconstruction of historical consciousness.

What was the transition from the Third Reich to the post-war period like when it came to one's relationship toward Jews? One partial answer to this is provided by a quotation from the recollections—freely narrated and in response to the interviewer's questions—of a woman born in 1923, who was active in the BDM from 1935, becoming a BDM leader and later a kindergarten teacher.

'Everything was wonderful', then came Kristallnacht, and soon 'it was over with'. She criticizes her father for not having told her the real truth after Kristallnacht. She mentions that after that she 'withdrew completely'. The symbolic function of the pogrom also appears repeatedly in other interviews. But did such a reorientation really take place in the mind of this 15-year-old girl? Later in the interview, she describes in glowing colour her experiences in the BDM: 'For me, the time I spent in the BDM was simply a wonderful period . . . Even back then, I'd often wonder about things, think that something wasn't quite, just wasn't quite right. Yet I simply rejected those thoughts by, by my inner attitude: what I'm doing is simply beautiful.' There was no change for her until 'shortly before the end of the war': not because her fiancé was killed in action, but because she was directly confronted with a dead soldier during a bombing raid. 'I passed out, and was unconscious for a long time; and then I thought to myself, "God, and this is what you lived for". And . . . that's when I started to reproach myself a lot.'

Toward the end of the war, she was working as director of a home for children. She recalls that the Americans came first, then the

Russians; in between there were people from the concentration camps. She and her mother had given the concentration camp people carrots to eat. But when she thought someone was Jewish, she gave them nothing—until her mother reprimanded her for such behaviour. Then she adds, unexpectedly, that after Kristallnacht her mother told her the Jews were being sent to Israel. And that the Jews had caused a lot of harm because of all the money they had. Then she mentions a Jewish family in which only the son survived.

Though her mind jumps back and forth in her narrative, it is striking that the connection with Kristallnacht is maintained. That is not surprising in view of her age at the time and the fact that this event, along with the truth about the concentration camps, was at the centre of attention in German and Allied publications after 1945—not Hitler's appointment as Chancellor in 1933 or the 1935 Nuremberg Laws and their after-effects. Her recollection of this date also coincides with her very intensive period of involvement in the BDM. However, she is not able to combine her supposed rejection of what was happening to 'those Jews over there' with the emotional description of the beautiful time she had had in the BDM. And at the end, she tells about encounters and contacts with 'girls from her group' of the BDM as late as 1981. This for her is an indication that she had in fact been a 'good leader' at the time.

In other recollections Jews appear as a moral element. The interviewees draw closer to them, yet immediately distance themselves once again, influenced by traditional stereotypes: because of their money, Jews had been the cause of a great deal of harm. Jews had been warned and had left town. What is striking here is often the difference in quality of emotion expressed in contrast to that in recollections of experiences in the BDM, the HJ, or even the SS.

The transition from the war to the post-war period and memories relating to Jews also play a role in other recollections. A woman from a wealthy family, born in 1930, and likewise a former member of the BDM, relates the following:

> We always used to play with the boy before then, and now we weren't even permitted to say hello to him any more. We were astonished that he was still around . . . My grandmother and grandfather had already joined the Party at the beginning of the 1930s. Simply because it was their opinion that everything would turn out OK then . . . Yet I can still remember that my grandma would always say: 'My God, I don't like having to walk past the K.s without saying a word to them.' Yet they didn't even look at you any more, they always turned

the other way when you went past them. R. and his mother were still receiving ration cards. But it was very astonishing and strange somehow that they were still around . . . On 5 May the French arrived, and the next day the neighbour of this family K. jumped off his balcony. And he was just an ordinary minor member of the party; nobody could understand that at the time.

So then we were walking down the street for the first time since the French had arrived, and R. was sitting on the bottom of the fence there. So we said: 'Hi, R.' And he replied: 'You don't need to say hello to me now. You didn't say hello to me for a long time.' 'Yes,' my grandmother said, 'and I'd like to apologize for our behaviour.' Then Mrs K. came along, and my grandmother went up to her and said: 'Mrs K., I'd like to apologize. Basically it's a terrible thing, what we did all these years . . . , and if I can help you in any way, Mrs K. . . .' Mrs K. replied: 'Let's just wait. Maybe in two weeks, I'll probably be able to help you.' And that's how it turned out.

At this point, one really ought to have the text of an interview with the young man R. sitting there on the fence. But, as in so many cases, even those Jews who stayed in Germany after liberation did not necessarily go on living in the very same surroundings where they had suffered so much. Many moved. Besides, the plight of the survivors was not finished in these early months after May 1945. There was not even a semblance of normality yet, as most German Jews living at this time in occupied Germany report in their narratives.

In some German narratives, Jews play a role only in regard to camps (and even here in a highly subordinated form) and in the memory of childhood experiences. A former SS man describes an episode involving a Jewish butcher who had been awarded the Medal of Honour and then had 'disappeared' in 1937, and whose whereabouts are completely unknown to him. When questioned, he recalled further details: 'Butcher Koppel had two big sons. Really typical Jews, big, burly guys, curly black hair, very dark types, really typically Jewish. I don't know though what became of them'[18]

It should be emphasized here once more that these interviews were conducted at the beginning of the 1980s. He describes a memory which clearly had gone through some kind of re-evaluation in the course of time. You can read the quote to mean that 'big burly guys' with 'curly black hair' are typical Jews. And you can interpret the second part to mean that such swarthy types are really typically Jewish.

More than anything else, what we have here is undoubtedly an exaggerated reversal (with the help of descriptive adjectives) of the anti-Semitic stereotype regarding the external appearance of the 'typical Jew'. The hunch-backed, hook-nosed figure with a devious look and black shaggy hair—as in the Grimm fairly-tales or, depending on one's age, the *Stürmer* newspaper—who supposedly sent fear and trembling down the spines of all 'Aryan' girls, is now supplanted by a new, positively exaggerated figure: big and strong. Perhaps this is even an exact and accurate description of the two sons of the butcher, yet the repeated intensity of the description is characteristic of stereotypical, philo-Semitic thinking about Jews. The reordering of bias shines through many interviews when it comes to this seemingly so touchy subject.

The following remarks, culled from various descriptions of situations, indicate that the above is not an isolated instance. A worker born in 1910 talks about Jewish merchants from whom you could buy products for very little money and on time, and then adds: 'and those were . . . such nice people', and they had 'such beautiful clothes'. In another description of the relationship with Jewish customers, a salesgirl born in 1922 talks about her impressions and perceptions. The propagandistic impact of the *Stürmer* is alluded to here, along with several of the contradictions characterizing transitions in the image of the Jew. They were rich, drove big cars—but the Jewish butcher was generous and gave things to the poor. The *Stürmer* painted a picture of ugly subhuman persons, but on closer inspection they turned out to be beautiful people, especially the girls. Both aspects—the packages of extra meat from the butcher and the beautiful Jewish girls—may have been a reality in this case, but both are in keeping with stereotypes associated with Jews that developed in the late 1940s and 1950s. At the same time, social stereotypes are presented which do not necessarily always have to be associated specifically with Jews. The perspective which singles out the possibilities of a more wealthy stratum in regard to appearance, clothing, and even hair-do naturally also exists quite independent of nationalist, racist, and anti-Semitic experience within the social experience of those individuals quoted. Ideals of physical beauty, wishes, and envy can also be involved. A relatively traditional package of prejudices regarding Jews is produced simply by being mixed with other stereotypes. (This, of course, is another challenge to the interpretation of oral sources.)

The same is also true of another memory: the Jewish doctor who was especially helpful to poverty-stricken female patients and did more than simply provide medical assistance itself. There is a hint of a philo-Semitic post-war filter operative in just such stories. It was through this filter mechanism that many memories relating to Jews were adjusted to conform with 'normal times'. Yet they are not pure figments of the imagination, or of a renewed indoctrination. They have their basis in specific individual and social experiences in the family, neighbourhood, and at work; these experiences underwent a social reinterpretation and re-evaluation in the developing political culture of the Federal Republic. They are important for memory, in telling one's life history, in the choice of words—less because they are largely in keeping with everyday reality under the Nazis than because they correspond to the later moral and cultural reality of the times.

The range of attitudes when discussion turns to surviving, returned, or supposedly returned Jews is more problematic. This extends from humanitarian statements to a reactualization of anti-Jewish clichés. A clerk from Essen born in 1916 notes, after he has described how life in his family 'developed normally' after the Nazi take-over:

> My wife knew a Jew, she helped him. After the war, they gave us a lot of assistance . . . Many people didn't want to believe the things about the concentration camps, but everyone knew . . . Our Jewish acquaintances were probably all sent to Theresienstadt, but they came back. My wife always helped them. I don't know the details though, since I was away most of the time.

Knowledge of concentration camps prior to 1945 is often admitted, but just as often, if not more frequently, awareness about this is postponed until after the end of the war. This is particularly striking if what is being narrated is not a memory of a personal experience. Nazi crimes are pushed aside, kept at a distance—before 1945, one had to learn, to work, to be a soldier; after 1945, one had 'to work in order to survive'. One man recalls:

> But, like I said, I didn't . . . hear about anyone who was persecuted because of his political or racial characteristics. I don't know anyone, so I never understood that either. It started up after the war, through the media, they said the Nazis had set fire to the Jews and burned them. Then we just said that they were crazy. Just like those stories about the hacked-off hands of children in the First World War . . . I only know that before the war there was a Jew here who dealt in junk. H. was his name. He had an old broken-down shop, a horse-drawn waggon and used to go around the neighbourhood collecting junk. He'd always come and pick up my grandfather's junk.

He goes on to mention that as a child he once heard his grandfather say to one of the other drivers: 'H. is really a very decent and honest Jew.' Then his grandfather was told that he was being cheated. At first, his grandfather did not want to believe this, but he had it checked out, and

> it was confirmed. Then the war started, and at some point the Jew was gone. I don't know, I didn't . . . directly notice when he disappeared, since I was no longer around here. In any event, after the war he showed up again. And got involved in junk again. He would go around picking up the junk again. Anyhow, those words [about cheating] were still running through my mind. So I followed him. It was exactly the same as before. When he understood that I had seen him, he came over and tried to pay me to keep quiet. I sat in the car together with this old man H. here in the yard. He had an old Borgward. Back then, nobody had a car, but he had this Borgward and a big five-ton Magirus-Deutz truck. He got this as part of restitution. Then he told me that he'd stood in front of the gas chambers a couple of times, with his entire family.

He also mentions H.'s remark about the pension the family was getting; that H. was sick and was convalescing 'in restitution'. And he sums up: 'He got a villa, a truck, a Borgward as restitution. And he's back again at his old business. Yes, . . . that's the reason people talked badly about him.'

The dividing line between the stereotypical impact before and after 1945 is very thin, though the restitution topic becomes obvious. With similar candour, this interviewee from Bochum, born in 1926, described his own legal and illegal activities with stolen goods and on the black market in the immediate post-war period. Such activities were all supposedly necessary for 'survival'. He begins his recollection by stating that he didn't know anyone who had been persecuted—yet goes on to tell his story about the Jewish junk dealer. The Jew vanished, but turned up again later on. Something kept running through the interviewee's mind: not the problem of persecution, not the question about where this Jewish family had returned from, but rather, whether this Jew was decent and honest or not. The fact that, in accordance with the story, he is not decent and honest, doesn't simply make him a dishonest junk dealer. The anti-Jewish economic stereotype underpins the entire narration. He's back again at his old business. For that reason, people talked badly about him.

Such traditional associations are linked here with restitution in the memory of the 1950s, though the episode recollected is indeed timeless. If it was shortly after the end of the war, the Jewish family would hardly have had large sums of restitution money available; if it was during the 1950s, he was hardly one of the 'very few' to have a car or truck. The observed dishonest behaviour of the old man H. is the thread on which traditional biases can be strung. Moreover, restitution can be included as an element in the economic stereotype of the rich Jew without asking about the underlying reason for such restitution. 'The government pays money for people who were not killed. I heard it's been established, proven that payment is being made for Jews who are still alive. But I had no interest in the details.'

It is important to realize that the interviews were given decades after the events related, so that particular political and cultural filters are obvious in the interviews. The way one *had* to talk about Jews in the course of the history of the Federal Republic also entails a certain way of narrative expression about matters related to Jews in the past.[19] Since anti-Semitism was tabooed by order of occupation authorities, and since a philo-Semitic mode of dealing with Jewish topics evolved in the late 1940s, one has to listen to the interviewees most carefully. Sometimes it is not what is being said but what is left out, which should arouse the interviewer's attention. Oral history is also the *art of silence*.

Attitudes toward Jews in many interviews cannot be categorized as either anti- or pro-Jewish. There is a spectrum of positions relating to Jews, ranging from critical-democratic views to philo-Semitic attitudes, and to highly distorted anti-Semitic biases—often found in the case of one and the same individual. The post-war condemnation of racial persecution and mass murder did not—by necessity—include a conscious rejection of stereotypes relating to Jews.

That leads us to another methodological problem when discussing Jewish topics in post-war German oral history, namely the way the interviewer handles Jewish issues or the participation—in the widest sense—of the interviewee in the persecution of the Jews. Sometimes interviews and reports have to be read very critically in order to decipher ways of 'being unable to listen' or even the rejection of offers by the interviewee to go into detail about the 'Jewish question'. The evasive manner in which the HJ generation deal with their involvement in the Shoah is sometimes complemented by the younger historians' selective evacuation of their historical consciousness by simply not asking the appropriate questions.[20] In a more comprehensive context, such problems are part of the 'disagreement in Germany, as to what belongs to the collective memory'.[21] It has to be stressed that this does not always include a conscious decision, though it does indicate the necessity for a historian to conceptualize the contemporary context of history and the moral implications of any interview which relates to the crimes perpetrated against the Jewish people. And this, of course, includes *every historical work* on the past decades of German history.

When dealing with political and social continuities, and with the *longue durée* of mentalities, one has to consider the continuities and metamorphosis of the image of the Jews in collective memory. This image in German historical consciousness, as articulated in many oral recollections, remains highly ambivalent in the period of transition from the Third Reich to the Federal Republic. Stereotypes changed and the yellow badge was partially whitewashed, yet the image remained stereotypical. The image of the Jew as perceived by many Germans is part of the continuous mental flow from the 1930s to the 1960s. The self-image of many Jews, especially those who stayed on in Germany or decided to leave, and their image of the Germans have never been and never will be the same after the unprecedented historical rupture. The 'alienation between two worlds of experience, of consciousness, and of collective memory' is a lasting and even growing phenomenon we have to confront.[22] The antagonistic experience of everyday life has led to antagonistic memories which have a decisive impact on historical consciousness.

Notes

1. Julie Heifetz, *Oral History and the Holocaust*, Oxford, 1985, 18.
2. See Meyer Levin, *In Search: An Autobiography*, New York, 1950, 228 f.
3. OMGUS, *Report on Conditions of Jews in Berlin* (15.9.1945), National Archives, Washington, Register Number 260, 44–45/6/9.
4. Levin, *In Search*, 232.
5. See Cilly Kugelmann, ' "Tell them in America we're still alive!": The Jewish community in the Federal Republic', *New German Critique*, 46 (Winter 1989), 129 f.
6. Rudolf Schottlaender, *Trotz allem ein Deutscher. Mein Lebensweg seit Jahrhundertbeginn*, Freiburg, Basle, and Vienna, 1986, 48 f.
7. Comité des Délégations Juives (ed.), *Das Schwarzbuch. Tatsachen und Dokumente: Die Lage der Juden in Deutschland 1933*, Paris, 1934, 9.
8. See Fritz Stern, *Dreams and Delusions: The Drama of German History*, New York, 1987, 97 ff.
9. Rudolf Lennert, 'Zugehörigkeit, Selbstbewusstsein, Fremdheit: Erinnerung an eine dunkle Zeit', *Neue Sammlung*, 3/26 (1986), 393.
10. Lotte Paepke, *Ich wurde vergessen: Bericht einer Jüdin, die das Dritte Reich uberlebte*, Freiburg, Basle, and Vienna, 1979, 104.
11. E. Ehrlich in *Jüdisches Gemeindeblatt für die britische Zone*, 5.4.1947.
12. Siegmund Weltlinger, 'Hast Du es schon vergessen? Erlebnisbericht aus der Zeit der Verfolgung', Vortrag in der Geselleschaft für Christlich–Jüdische Zusammenarbeit, Berlin, 28.1.1954.
13. Paepke, *Ich wurde vergessen*, 120 ff.
14. Marga Spiegel, *Retter in der Nacht: Wie eine jüdische Familie überlebte*, Cologne, 1987, 72 f.

15. The following discussion is based on materials from the project 'Lebensgeschichte und Sozialstruktur im Ruhrgebiet 1930 bis 1960' (LUSIR). The interviews were conducted in the Ruhr (in Duisburg, Mülheim an der Ruhr, Essen, Bottrop, Gelsenkirchen, Recklinghausen, and Oer-Erckenschwick) between 1980 and 1982. In the framework of the interviews, questions were asked, among other things, about whether the respondents knew any Jewish families, whether they had had Jewish schoolmates, and whether they could recall their whereabouts and remembered people wearing the yellow star. Recollections of the Reichskristallnacht, of burning synagogues, of the persecution of Jews, and of Nazi racial ideology were dealt with as themes. Since these interviews were conducted in the framework of the main focus of the project, the relationship between Germans and Jews was not a special theme singled out for investigation. For this reason, there were no follow-up questions relating to this subject. The only interviews utilized here were those which proved useful for the topic in hand (40 interviews). These interviews could not have been utilized without the assistance of Lutz Niethammer, Nori Möding, and Alexander von Plato. The author wishes to express his gratitude to them for this assistance and assumes, of course, full responsibility for the presentation and interpretation of the material. An extensive analysis of these interviews is given in the author's thesis (see the footnote at the beginning of the chapter).
16. See Sybille Hübner-Funk, 'Jugend als Symbol des politischen Neubeginns: Strategien zun Bannung der rassistischen Vergangenheit', in Werner Bergmann and Rainer Erb (eds.), *Antisemitismus in der Politischen Kultur nach 1945*, Opladen, 1990, 218 ff.; Heinz Bude, *Deutsche Karrieren: Lebenskonstruktionen sozialer Aufsteiger aus der Flakhelfer-Generation*, Frankfurt, 1987; Gabriele Rosenthal (ed.), *Die Hitlerjugend-Generation: Biographische Thematisierung als Vergangenheitsbewältigung*, Essen, 1986; Heilwig von der Mehden, *Der Friede sah ganz anders aus: Junge Menschen 1947*, Freiburg, 1984.
17. On '*Habitus*' see Pierre Bourdieu, *Zur Soziologie der symbolischen Formen*, Frankfurt, 1974.
18. This and the following quotations are taken from LUSIR.
19. See Frank Stern, 'Philosemitism', *Holocaust and Genocide Studies*, 4/4 (1990) 463 ff.
20. See the depiction of this problem by Gabriele Rosenthal, 'Leben mit der NS-Vergangenheit heute. Zur Reparatur einer fragwürdigen Vergangenheit', in *Vorgänge* (May 1989), 97 f.: 'What is involved here is the mutual construction of protective walls as a product of interaction, in which the narrator discloses only parts and retains others hidden behind a wall and the listener wards off partial memories by constructing a second wall' (p. 98).
21. Dan Bar-On, *Legacy of Silence: Encounters with Children of the Third Reich*, Cambridge, 1989, 328.
22. See Selma Leydesdorff, 'Das gebrochene Schweigen: Lebensgeschichten von Überlebenden des jüdischen Proletariats in Amsterdam', in *Bios. Zeitschrift für Biographieforschung und Oral History*, 2 (1988), 25.

3

Where Were *You* on 17 June?

A Niche in Memory

LUTZ NIETHAMMER

Within the countries of the former Soviet bloc, the date of 17 June 1953 stands out as *the* key event in the history of the German Democratic Republic (GDR): in this context, it represents the first broad popular uprising against Stalinism after Stalin's death. Yet if one compares this brief, one-to-two-day protest in the GDR with events in Hungary and Poland in 1956, Czechoslovakia in 1968, and then with Poland in the 1980s—and, in a less spectacular form, with changes in Hungary as well—the character of that 1953 protest remains indistinct and controversial. Nevertheless, however short-lived, it was a forerunner. The unrest marked a watershed in the history of the GDR, one that would not be repeated until its disintegration in the autumn of 1989: a spontaneous protest that spread like wildfire during the course of a single day to most cities and industrial centres throughout the country, at the same time questioning the very existence of the political order there. From the perspective of its Eastern bloc neighbours, the characteristic features of this primal conflict were the precocity of its outbreak, the threat it posed to state order, and its non-repeatability. Over the following thirty-six years—and thus throughout a period longer than in any other of the industrialized avowedly-socialist countries, the GDR was able to dampen and contain social conflict, although at the cost of a spectacular degree of self-isolation within its own nation in the wake of the wall's construction. *Solidarność* remained a distinctly foreign word.

Public perception of this basic conflict was highly politicized right from the start and bound up with the question of the existence and survival of the GDR as a state. The leadership of the ruling Socialist Unity Party (SED) denounced the events of 17 June as an attempted coup instigated by fascists and Western imperialists. It was officially

Translator: Bill Temple.

contended that popular sentiment in favour of the coup had been fuelled by 'errors' of the GDR leadership in pressing ahead too rapidly with the building of socialism. After quashing the uprising at the hands of the Red Army and effecting a slow-down in the excessive tempo of socialization, the gap that had opened up between the party and the people was closed once again. In the media and among historians, a concrete perception of that day's events was rendered virtually impossible: the files vanished. People officially in the know cited a figure of 6 per cent for participation by the population and were careful to add that it also included former Nazis in the GDR. Many of those who took part, both speakers of the protest movement and representatives of institutions and the SED who had 'weakened', were persecuted or dismissed. A number avoided this fate by fleeing to the West. This lingering trace of persecution and taboo, along with the continuity of the regime and the West German reaction, all contributed to the official legend remaining intact, fixed, and unamenable to revision.

The prevailing assessment of the events of 17 June in West Germany, framed in national and anti-Communist categories, was that it had been a broadly based 'popular uprising' against the Soviet Union and its minions in the 'Eastern Zone'. Its proclamation as a national day of remembrance functioned to underscore the claim of the Bonn government at the time that it was the sole and exclusive representative of all Germans. Yet that interpretation already began to erode a decade later in the wake of the growing readiness in West Germany to accept the *fait accompli* of two states, after the building of the wall. On the one hand, public interest waned to the extent that a sense of West German identity and belongingness crystallized. Moreover, with the advent of the new *Ostpolitik*, official celebration of 17 June increasingly became a kind of public embarrassment and calamity. On the other hand, research on contemporary history had plumbed sources available in the West and advanced an alternative reading of 17 June: it was viewed as a 'protest by workers', and thus, at least to a certain extent, as an internal problem of the GDR. In this connection, such research could point not only to the predominance of workers in the demonstrations and strikes, but also to the apparent triggering factor: namely, discriminatory treatment of the workers. After the revocation of the excessive measures against the middle class, initially introduced as part of the new governmental strategy, the simultaneously instituted increases in norms for workers had nonetheless been retained.

Most recently, in the wake of the proven inability of the SED regime to integrate itself into the new Soviet policy toward Europe—initiated by Gorbachev—and the regime's apparent incapacity to maintain itself independently, the phase of two German states has come to an end with the expansion of the Federal Republic eastward to the Oder. The internal factors that gave impetus to this process of unification, namely the stream of youthful immigrants flocking from the GDR to the West in the second half of 1989 and the shift in the demonstrations that took place in the autumn of 1989—from protests for liberalizing the GDR to a national push for unification with the Federal Republic—not only brought about the demise of the GDR; it also reopened discussion about the interpretation of 17 June as the primal conflict of the GDR and of the analogy between 1953 and 1989 in the dynamics of their underlying factors. Moreover, the domestic (ex-)GDR sources, formerly inaccessible, can now also be examined.

Was it really a popular uprising? And if so, what is the significance of that fact? Put differently: is it possible that the GDR never really found a sense of its own place in historical time or its own realm of experience, but rather was, right from the start, caught up in the time-jam of an unresolvable dilemma? In that dilemma it was immaterial whether people said 'we are the people' a few years after the GDR's beginning or a few months before its end—because, in any event, after some days or several months, the slogan that in fact resounded was: 'We are one people!' The history of the GDR lies strung between these endpoints: Was it merely a moment frozen in time, a warp in which nothing really changed and the only thing that aged was the population? Or was it a different people after all, one that eight years after the fall of the Third Reich had gone into the streets to protest against Ulbricht?

If one takes another look at the GDR, and compares it with its neighbours to the east and south, then it can arguably be viewed as a society where conflict was repressed and dampened for a period extending from the middle of the 1950s into the mid-1980s. It was able to function in this way because, compared with the general prevailing economic situation in the Eastern bloc, the GDR offered a relatively large number of social opportunities and served as the showcase model of the socialist camp and as the junior partner of the Soviet Union within that band of poorly developed societies less oriented toward output and achievement and/or beset by serious internal conflicts. Now, after having moved from the best floor in the socialist poorhouse to the attic of the capitalist department store, a method of creating tradition may perhaps suggest itself: an approach that pastes one Western option onto the next, rendering one's own history nugatory in the process. Yet human experience is not liberated in that way.

As an external observer who cannot possibly anticipate the findings of future empirical research on the fundamental primal conflict of the GDR, I do not wish to pretend that I, a historian based in the West, already know what *really* happened in the GDR in June 1953. Nonetheless, I would like to present some findings derived from the evaluation of life-history interviews that our research team conducted and recorded in a number of industrial centres in the GDR in 1987.[1] In these broad-ranging discussions, we also gathered data on the question: Where were you on 17 June, what were you doing on that day? I proceed on the basis of the hypothesis that experiences at that time preshaped—or, to put it more accurately, restricted future perceptions and spheres of action. That is, that the processing in memory undergone by 17 June can tell us something about the political culture in the GDR as a whole. And I would like to place this processing, at least in part, within the context of life histories before and after 1953 in order to contribute to an explanation of the riddle why the GDR developed after June 1953 along more successful lines than, say, Czechoslovakia, and yet appeared to have less internal conflict than a socialist sister-state like Romania.

Allow me one more prefatory comment of admonition: my formulation of the problem may not find favour with a great many people in the former GDR who are now torn between feelings of being showered with gifts by the West, while at the same time being expropriated: they may see in me just one more 'Wessie' who pretends to know more about them than they know about themselves. But they can rest assured that I have no such pretensions. Rather, I only wish to raise certain questions, and do that principally because of the nature of the data available to us: so many persons reported to me—and to my co-researchers Alexander von Plato and Dorothee Wierling—about their lives in the GDR, and did so in more detailed interviews than even GDR citizens had ever been permitted to conduct under the *ancien régime*.

A first initial observation can lead us into the heart of the central problem, which involves the extent to which one's own experience can be linked with collective memory: significantly, the only persons among our interviewees, who were all aged between eighteen and fifty-five in 1953, the only persons who brought up the topic of their experiences on 17 June were members of the SED, in particular functionaries. Evidently, the official legend of the attempted imperialist coup and the misguided masses gave them enough support and reinforcement so that they also felt they could venture a report on their own experiences and were not afraid to talk about this topic, generally recognized to be a controversial and conflict-loaded subject. In contrast, none of those who had been a direct participant in the demonstrations, or who had encountered them from the sidelines as a politically non-involved spectator, had this official legend to fall back on for support. Only after asking specific questions were we able to learn something about the events in question, and even then, not in all instances. The reason for this reticence was not that 17 June was unfamiliar or had little or no importance to the specific interviewee—all of those we interviewed, even if they had difficulty remembering dates, knew immediately what we were asking about when they heard '17 June'. Only once did someone inquire as to what year we were referring to, and that question came from a 26-year-old woman who was accidentally present during an interview with her grandfather—much to his chagrin. The construction of the wall and 17 June remain the most vivid dates in GDR history for those generations that directly experienced the events. The vivid presence of those dates was bolstered by the 'external' collective memory, as it were, of the Western media, though memory inside the GDR pointed to the existence of an intimate private sphere to which one could gain access by asking specific questions and acquiring the interviewee's confidence.

There are basically three types of reaction to such inquiries. First, a large number of those interviewed stated that they had had no special experiences on that particular day. In most instances the explanation given was that at the time in question they had been in some rural area or in the south-east of the GDR, where it was impossible to pick up the programme of the RIAS radio station in West Berlin. That is, they claimed that they had heard about the events, limited to a brief period of two days in Berlin and merely a single day in other industrial centres, only after everything was over and finished. On a number of occasions, informants also assured us that they were unable to remember anything, since there was as yet no TV back in those days. However, that statement about political perception tended to be more characteristic of the passive practice of the later GDR, saturated by Western media, than of the earlier years of the republic, significantly shaped by radio and word-of-mouth information networks. One variant of non-reply, suggesting an excessively intensive degree of experience rather than one too meagre, was the open refusal on the part of informants to answer. Thus, for example, there was this telling reaction of a steel-mill worker: after the question had been raised in a joint interview, he got up, smiled, and, excusing himself repeatedly, stated that he had to go to the toilet; upon returning, it proved impossible to manœuvre him back into dealing with the question. Or take the instance of an electrician in Bitterfeld who, when asked why he wasn't a member of the SED and whether he had had any disadvantages because of that, answered, punctuating his reply by several sighs, that those were questions which simply went too far:

What did you do on 17 June?

Well, I was with some friends here in the market.

Ah, you were here. Was there a lot going on?

Yes.

And why did you join in?

Well, I'd rather not go into that, since I left the window open at the back. Let's just leave that question. No.

The relatively small number of interviewees who admitted to taking an active part in the strikes or demonstrations—the basic content of their memories will be dealt with below—did so in a later section of the interview, after a certain relationship of trust had been established and we had reiterated that we could take our tapes out of the GDR without their being checked by government censors. Moreover, they were not particularly worried about possible sanctions against them should the information they gave be misused, because two of them were already very advanced in age anyhow, and two were very young or unemployed. The fifth person had had a few drinks before the interview to build up his courage—similar to an experience of his in 1948 when, in a drunken stupor at a factory celebration, he had joined the SED, and then afterwards had not got up the gumption to resign from the party. This informant had served in an élite unit during the Third Reich; his father had been an active Nazi who had done well pursuing a career in the Warthegau (occupied Poland), subsequently committing suicide in 1945 when these territories were returned to Polish control. Now he was a skilled, highly respected worker, on the threshold of early retirement, his health ruined. Three of the four other men also enjoyed a high level of social respect in the GDR, and their previous biographies had all been connected, directly or indirectly, with National Socialism. One was a retired director of a commercial department in a large factory; despite his former membership of the Nazi Party, his veneration for the Prussian officer corps, and his anti-Semitic views, he had not been subjected to denazification in the Soviet Occupation Zone, but had continued in his job. In 1954, he was actually given a promotion, and offered membership of the SED, which he declined. His children ranged from an SED functionary to the wife of a pastor. The second was a professional gardener and active Christian who was respected in his parish for his level of education and thoughtful manner. He had been influenced by the youth movement and the Hitler Youth (HJ), and still organized a yearly reunion with several dozen—now greying—former buddies from the HJ. The third informant had been an apprentice in 1953 after his father, a Nazi, had gone to the West in 1950 following the death of his wife, abandoning his two teenage sons back in the GDR. He later became the director of apprentice training in a large factory, and after passing his master-craftsman exams, joined the Liberal Democratic Party (LDPD), but was not an active member. His interests concentrated on the job of being an educator and the building of a quite impressive bungalow. Finally, our group included a cleaning-woman who had always been in financial straits; on 17 June, she had joined the marchers after noticing from the sidelines that her husband was among the demonstrators. He was an office messenger who had been in the Nazi Party and the Nazi storm-troops (SA), and had returned from the war with serious injuries.

This biographical survey could give the impression to some that I would in fact like to muster support for the SED line of interpretation which views the events of 17 June as a fascist coup attempt. Nothing could be further from the truth. All I wish to do is characterize the background of our 1987 interviewees in the GDR who openly admitted to having been actively involved in the 17 June unrest. It is known, however, that in the city where most of our informants came from, and which was a former bastion of the left-wing Labour movement, the especially explosive turn that events took on 17 June was initiated and carried out by the great mass of the skilled work-force from several large industrial plants. Moreover, the strike committee, chosen by acclamation at a meeting of some 50,000, did not include a single Nazi, but did have a few Social Democrats and one Communist on it. Most of the strike leaders migrated to the West after the events; only a minority stayed on, and they did not long survive their period of incarceration, which in some cases lasted for many years. In any event, it proved impossible to locate any of those who had played central roles in the strike; those who now said they had taken an active part in the demonstrations had not been part of the leadership nucleus. The fact that they did candidly admit to involvement appears to be bound up with another circumstance: namely that they no longer had any desire (or ability) to get ahead socially, and had remained politically marginal in the GDR due to their past in the Third Reich. Apparently for this reason, they found it easier to cast caution aside in speaking with a West German. In other words: memory of active participation on 17 June had remained private. Unshared, it had not developed any powerful impact or influence on others.

The third type of reaction among our interviewees who did not belong to the SED was not only much more frequent, but also seems to me to be especially characteristic of the fragmented mode of memory in GDR society. What is involved here are what could be called 'choreographies of informed non-participation'. Already in the 1950s, this type of information was regarded by analysts (such as Lewis Edinger) as a characteristic legacy left by totalitarianism to democracy, entailing a mixture of a high degree of information combined with low affectivity and participation levels. In our investigation, it produced several curious accounts.

Thus, for example, the director of an experimental nursery in a chemical plant replied as follows to my question about what he had been doing on 17 June: 'Naturally, that's something that has to do once again with bees.' He kept bees as a hobby and had gone by bike on 17 June to Delitzsch to get a hive nucleus from a friend. Even in Delitzsch, everyone was out in the street, but he, his small swarm of bees on his back, had to hurry home, sailing on right through the crowd. He can thus depict in a credible way how he was able to know what was happening, and yet why he had not taken part—because of his bees. The pastor of Fürstenberg, otherwise a warrior in matters of the Lord, only noticed quite briefly in passing that construction workers down the block had emptied the town hall and torn down the socialist banners, and that then some tanks had rolled up. He had noted all this because every Wednesday, early in the evening, it was his custom to go over to a friend's place, a doctor, to play bridge. And he was still able to recall the specific day, because that particular Wednesday evening he had had to turn around and go home, due to the fact that a curfew had been declared.

A chief bookkeeper from the recently renamed city of Stalinstadt tells the tale of seeing the demonstration of workers from the biggest construction site in the Republic, standing with their banners near the train station in the afternoon: 'They were happy that now maybe there would be an end to the occupation, that they could get rid of the Russians. But it didn't even take half an hour and the tanks were there, and all that was smashed to bits. In an hour, everything was over with.' Responding with special sensitivity to this projective interpretation—because the withdrawal of the Soviets was rarely voiced as a demand in the strikes, and it is known that in that particular town, there were only demonstrations in the morning and at noon, while the Russian tanks did not arrive until the evening—the interviewer asked her why she didn't join the procession. She replied: 'Well, 'cause I was waiting for my train.'

A mason had already heard that morning that fellow-workers were throwing down their tools. This man had been a farm-hand in East Prussia and a member of the SA; now, despite expulsion, he held a good position with a future. He was for maintaining order; anyhow, he could not do anything without the rest of his work brigade. He claims he did not strike, but did not work either; instead, he got a bottle of wine from the plant canteen, sat down on a crate in the yard and waited. A head dispatcher at the freight station in Bitterfeld also noticed quite early on that workers from the plants were shutting everything down. He was an old Social Democrat, but had still managed to survive somehow in 1933 and in 1945 as a civil servant. He mentions grabbing a couple of bottles of beer, going to his vegetable garden, and whiling away the whole day there.

Even more distant, but likewise more informative, was the observation post of a chemical worker who had been in the Waffen–SS. 'Thank God', he had been given a day off in the middle of the week for having worked on Sunday, and so was with his mother in the fields mowing hay.

It was a good thing that I was down there in the meadow . . . And suddenly some people showed up, already coming back from work. I asked: 'What's going on today?' 'Oh, haven't you heard?' I said: 'No.' 'Hey, they're on strike down there. The whole plant's on strike.' So I said: 'What:!' 'Yeah, what do you think?' Then I heard about everything. And the second day I went on down. First I looked around to see what was happening. Yeah, everything was quiet, shut down. And then they came with the Russians and shot into the air.

The plant, which usually worked three shifts, stood idle for three days. Not until Sunday did they start working again. He adds that afterwards a lot of things changed, 'in part for the better'. After that, the best years of all came, a time when you could buy everything you wanted.

What acts to favour the construction of such removed observation posts on events, cautious middle positions, and passages of distanced spectatorship and witness testimony in memory? Initially, one might think that this is really simply what happened, and in individual cases, such as the last one above, this might indeed be a sufficient explanation. In the majority of instances, reference to the fleetingness and distance of perception serves to mask two elements. First, one's personal involvement and assessment, and second, the fact that the actual perception must have been far more complex than the particular perspective on which the observers settled and which they established in the niche of their memory—a perspective in which they are rendered unassailable by dint of their spatial position alone. At the same time, taking a personal stand of one's own becomes superfluous; its content can now be recast and poured into the form of a report about other people.

This handy way out via the passive route suggested itself for several reasons, and was also in keeping with the ambivalence of the reported phenomenon. Only a few hours lay between the outbreak and collapse of the protest movement, and it did not disintegrate as the result of a battle with armed force; rather, the mere appearance of that armoured force had restored the old order. In many instances, the crowds had already dispersed, and only in individual cases did people note that Soviet soldiers were patrolling the streets or that tanks had been driven up into position. Clashes with troops, as attested for Berlin, are not mentioned in a single one of our protocols gathered in the provinces. Moreover, the outbreak of protest had not been preceded by any social process of clarification of the issue, so that the argument advanced by the SED, namely that many had jumped on the movement's bandwagon without really knowing what it was all about, may indeed have had some basis in fact. In any case, in retrospect it offered numerous individuals a path for retreat—in particular, SED members who had joined the demonstrators mentioned this depressing argument which was used for their later rehabilitation: that they were guilty of having hopped blindly onto the bandwagon. Finally, political control mechanisms and industrial discipline appeared to have been relaxed in the days and weeks following the events, until they were stiffened once more a few months later: in factories, discussions were organized and grievance committees established, and these included participation by strikers as well. The material improvements brought by the 'new course' were implemented and increases in norm levels revoked.

Once the general political implications of the protest had been blocked by the symbolic presence of armed force and the protest reduced to its economic reasons, it was possible for the unrest to take on the appearance of an issue that had been resolved and eliminated by the subsequent material improvements. The chemical worker in the hayfield, who recalls the later 1950s as the most beautiful years of his life, because back then you could already (and still) buy everything you could afford, was certainly not an isolated instance among the older generation in the GDR. As a result of the lack of political alternatives, the masses fell back into passivity and economism. The content of protest issues, which, once the movement had gained momentum, were intensified within a very short period, and heightened into the demand for the overthrow of the government, free elections, and an expression of national feeling, had not only been repressed by public censorship but also lost their salience in private memory. What remained in the niche of personal memory was only an image of that autonomous movement that had shattered the bounds of the permissible and whose dynamics had given expression to repressed sentiments. Because of its failure and stigmatization by the old–new regime, it became necessary to distance oneself from what had transpired. Yet, the image of its unanticipated possibility remained preserved in memory—though, of course, not as *one's own* possibility.

As already mentioned, it was not the participants in or witnesses to the events of 17 June who, on their own initiative, continued to mould and shape the image of its tradition in the GDR, but rather the opponents of those events, the SED functionaries. There are two reasons why they were the only ones among those interviewed who spontaneously volunteered to speak about these events or, if they had not mentioned them, readily answered our probing questions. In addition to the reason already mentioned, namely that they were able to find support for their position in the official collective memory of the GDR, there was another, contrary one: they felt they were on the defensive against the 'second public sphere' of the GDR, in which the Western media attempted to link up with the word-of-mouth network of non-official popular opinion in the GDR. This was especially the case because the Bonn government had made 17 June into an official state holiday; hence, many SED members tended to overestimate its importance to the West German citizenry. In any event, several interviewees indicated that they regarded the holiday as a kind of official state mourning for the failure of the so-called 'Day X', on which the Federal Republic had allegedly hoped to bring about the collapse of the GDR. Though it should be added that this 'Day X' was known as a concept only in official propaganda in the GDR, and was not familiar to West Germans.

In contrast with the three types of remembrance of 17 June discussed above, the SED members constituted a further type; this could be associated with (or displaced by) two subtypes. The main type in this instance involved a personal certification of the official legend of the fascist or imperialist coup attempt and the gullibility of the people who were led astray and duped by the rabble-rousers. The first subvariant stressed that the narrator himself had been among the duped and had joined in this demonstration, as in many others, only to discover too late that it was not an official demonstration, and in effect was directed against the regime itself. The second subvariant presents 17 June as a kind of 'situation of awakening', in which mainly younger people became aware that the 'people's property' was endangered or threatened by acts of sabotage. They then chose to become 'more socially active', as a rule by joining the party or the factory combat units, used to inculcate military discipline in nuclear groups of workers in the wake of 17 June.

The problem associated with the main type of SED remembering was that difficulties arose when it came to exemplifying the party legend in one's own experience, especially in talking about how the narrator had behaved and experienced the masses of workers. As a result, the informant often either lost control of the thread of the story, or the report had to be so drastically abbreviated that it forfeited all credibility. To this extent, collective memory existed only on paper: when memory had to be articulated, it was undermined by personal recall, undercut by the niches in the memory of the functionaries themselves.

The ritual demand of *Bewährung*—'proving one's worth' and loyalty—entailed declaring one's allegiance to the party in a critical situation as well, and giving expression to that declaration in exemplary behaviour *vis-à-vis* the masses. Yet, that was either totally ineffectual or fraught with risk. Thus, the leading party functionary in Bitterfeld was thrown into the nearest river by the crowd, during such an attempt; the head of a union in a chemical plant had to watch in disbelief as his supposed fellow-workers failed to heed his warnings and marched on past him in rows of eight men deep; a youthful functionary, dressed in the uniform of Dienst für Deutschland, a voluntary service branch of the Free German Youth (FDJ), soon found himself standing in nothing but his underpants out on the street. It is not surprising that the head of personnel in the municipal administration, when on his way back from a small factory—the only one not on strike in town because pay-day had been arranged for 17 June—to Party headquarters—just being ransacked by demonstrators and their contents tossed out of the window—quickly grabbed the arm of a faithful female activist; disguised as a 'pair of lovers', they made their way unnoticed through the 'mob'. A later director of a factory union presented as proof of his worth and loyalty the story that he had not gone on strike, but had continued to work on that day of 'testing'. When questioned, it subsequently turned out that, not yet in the party at the time, he was on duty at the plant switchboard and that a representative of the strike committee had specifically told him to stay at his post. A female party functionary in Halle, who was taking care of a baby at home, wanted to prove her worth and loyalty by not removing her party insignia during a walk with her child to the polyclinic in the centre of town. Yet, already a few blocks before her destination, she had been forced to turn back when faced with the large crowds, and she told how she had taken the baby out of the buggy and pressed it to her bosom. When questioned, it turned out that she had done this not because of any threat to the child, but because she had been resistant to taking off her party badge. An FDJ activist—formerly a member of the Nazi girls' movement Bund Deutscher Mädel (BDM)—who had run from the plant early that morning when faced with the spectacle of 'a few threatening and angry faces in the neighbourhood', placed herself at the disposal of her party leadership, which for its part had retreated for cover to the administrative wing of a state factory. Her 'proof of worth' consisted of the following: she, and apprentices from the dormitory she was in charge of, proceeded to prepare sandwiches and coffee for the plant managers who were in hiding. She was later promoted, rising to the position of director of the plant, and emphasizes that after this incident she had been able to write her apprentices letters of reference for their party file attesting to their loyalty—recommendations that proved extremely useful for their later careers.

Indeed, for the entire area of the chemical industry in East Germany, we are aware of only one successful 'proof of worth' that did not consist of a disguised retreat, night patrol to 'protect buildings' after everything was over, or contact with the Soviet troop camp looking for assistance. This instance of *Bewährung* was a counter-demonstration. It was narrated by an old Social Democrat, who, as one of the founding members of his local SED, had already been squeezed out of the ranks of the party leadership of his factory by the Communists in 1950, but who had remained loyal after being assigned the post of janitor in the party school. When the procession of demonstrators, led by a master craftsman, former Social Democrat and shop steward, passed by the school and expressed its desire to rip down the posters there, he stepped out of the front door accompanied by his dogs—a German shepherd and a large watchdog—and threatened to let them off the leash. So the posters stayed—they announced the recent rechristening of the town where the large combine Iron and Steel Works East was located to the new name 'Stalinstadt'.

Such contrary courage was a rare occurrence. Rather, the reports are permeated with the motifs of anxiety and surprise already alluded to. Generally, it is openly admitted that the grass roots movement came totally unexpected and 'like a bolt from the blue', but this is not accompanied by any reflection whatsoever about how it was possible for the bureaucracy to have been caught so completely off-guard, and why the workers and the force supposedly leading them had faced each other in such deep alienation. Rather, in individual cases (like that of a cadre leader) expressions are used such as 'mob' and 'hooligans'—implying that weapons should have been distributed to the functionaries for use against them. In the majority of cases, conspiracy theories are mentioned involving Western or fascist operatives behind the scenes, though these are never referred to by specific name (while the strike spokespersons mentioned by name generally were former Social Democrats known in the locality).

Finally, the alienating distance from the base is often frozen in the code of the clumsy and revealing language of the functionaries: for example, it was stated that 'back then', there were still 'a large percentage of workers who had not been schooled politically and who were confused in practical terms'. Or take the example of a baker's daughter, eager to acquire a better education, who had taken over her father's business from her older brother and had joined the party in early 1953 in order to acquire the proper qualifications for the post of editor of a plant newspaper. She expressed her astonishment at workers rising up against their own state, attributing this to the workers' lack of education: 'A Communist, well, he really has to be a very well-educated person, and we don't have so many people like that yet. To be honest with you—I don't think I'll even be alive then [when we do].' Or an objection is raised that changes in the party line are repeatedly necessary, but to force them by strikes and demonstrations is 'totally contrary to our thinking'.

This statement was made by one of the few women in the top echelon of industrial management in the GDR, the director of a large industrial complex and the daughter of a worker. During the war, she attended a National Socialist teachers' seminary. In June 1952, she was the wife of the head of the local party school; on 17 June, despite the fact (or because) she was in her eighth month of pregnancy, she was sent in an official car to Berlin bearing confidential documents for the threatened central office. Yet, by the time they reached the outlying districts of Berlin, it was clear that it would be impossible to continue on in an official vehicle, and she had to carry out her assigned task—her documents hidden under the cloak of her advanced pregnancy—by striking off on a gruelling march on foot. The chauffeur who accompanied her kept urging her on, telling her that her destination was just around the corner. It is clear that her memory of 17 June is of 'an enormous physical effort', especially when she adds that it was 'a day when the heat was scorching'. Yet, the Berlin weather report indicates rain for the 17th—it was hot in another sense. The translation of political preceptions into meteorological metaphors like 'prevailing weather system' or 'thaw in the weather' has a special tradition in the GDR. It characterizes grand politics as a domain which is both predetermined and variable, yet which cannot be shaped by human efforts: after all, one has to accommodate oneself to the weather. Those who had acclimatized themselves to the Cold War were unprepared for a sudden heatwave. Comparable projective metaphors for the feelings of fear and existential insecurity that the unbridled masses unleashed among functionaries can be found throughout their memories of that surprising day in June.

To give another example: the cadre leader, who bemoaned the fact that no weapons had been distributed to the functionaries, mentioned on the first day of his interview that his wife's life had been threatened then. In the further course of the conversation, with his wife present, he corrected this by saying that she had been threatened by the 'loudmouth "unity" hooligans'. When his wife told her version of what had occurred, it turned out that she had been at home with her small children, and had felt threatened when the procession of demonstrators passed through their street; she had locked herself in the apartment after she thought someone was 'fumbling' at the door. When she was unable to recall any threatening shouts from the street and only stated 'You can see from the way someone looks whether he's friendly or threatening', her husband—who at that moment had been standing (as described earlier) in front of party headquarters, disguised in a lover's embrace in the midst of the 'rioting crowd'—offered the following remark as a kind of prop to explain her anxiety out in their housing development: 'They were shouting very loudly: "The time has come, we have to settle accounts with you people!" ' Yet, she added a further comment, for the sake of accuracy: 'There were some who heard them shouting that. But I was so frightened I didn't hear it myself.'

By the decoding of such metaphors and projections, one arrives at another level in the experiential history of 17 June: the conflict of present feelings and those that had preceded them. In order to localize this clash correctly in historical terms, I would like first, by way of a brief digression, to present an observation and to report on a provocative assessment offered by one of our informants.

First the observation: in our interviews, conducted predominantly with workers, one can note that the more detailed and emotional reports about 17 June seldom came from workers; rather, they were given above all by members of the middle classes and functionaries. Now it would certainly be a mistake to regard this as an indication that the strikes and demonstrations were not initiated and carried out in the main by the workers. This is refuted by most of the data previously available, the outbreak of strike actions as a response to the question of increased norms, as well as the slogan chanted by the marchers, for example. Workers were pouring out of such a large number of plants and moving in the direction of Bitterfeld that a functionary, who had sought refuge on a roof, ironically said—parodying a party line about the 'Bitterfeld path' (in socialist literature)—that he didn't realize so many paths led to Bitterfeld. The slogan they chanted was: 'We don't want to be slaves any longer. Fellow-workers, join the march!' ('Wir wollen nicht mehr Sklaven sein, Kollegen reiht Euch ein!'). That was a slogan imbued with the spirit of the old workers' movement; it was a reaction not just to Stalinism, but to Fascism and to the oppression of workers everywhere. Rather, the conclusion that should be drawn is, that although a strike of this magnitude would be unheard of in the subsequent history of the GDR, it was not such a singular and isolated event in the previous experience of the workers that it might have been etched into their memories as a kind of primal event.

Next, let me call attention to what I regard as a provocative assessment by a former Social Democrat; quite different from the anxious projections of many of his colleagues in the SED. This person had been fired twice without notice during the Weimar Republic and the Third Reich due to so-called 'defiant behaviour' on the job; he was the man who, together with his dogs, had protected the Stalin posters from the demonstrators. He termed 17 June a 'minor affair', and compared it to a family quarrel that is soon forgotten, pointing by way of contrast to the often long-term labour struggles in many parts of the world, disputes that occasionally develop into pitched battles reminiscent of civil war. Measured by the criteria of a labour struggle, 17 June was only something like a spontaneous warning strike, very brief, broadly based, and extremely effective—in fact, in relation to its length, presumably the most successful such strike ever staged. The original economic trigger of the strike, the hike in norm levels (quite justifiable from the standpoint of productivity), was completely revoked, and most of the employers, who had been recruited from the workers' own ranks, had kept a low profile for months due to their fears. They had also learned the long-term lesson that one shouldn't tamper with the workers' pay envelope for political reasons.

But in another and more profound sense, 17 June was a total failure, because it should not be assessed and measured by the yardstick of a labour struggle only. In this regard, it was not a struggle at all, or even an uprising, but a conflict of feelings that erupted on one specific day and then lay buried for a long time to come. On the one hand, there was the shock to the SED cadre, who out in the provinces were anything but big-time party bosses, of realizing that they had completely lost contact with their fellow-workers—a work-force for whom they wished to construct a new order. The illusion of authoritarian normality had robbed them of any instinct for sensing what was actually taking place in the heads of the workers. They believed the outbreak of protest to have come 'like a bolt from the blue', and they had no other explanation than that there must have been dark forces at work behind the scenes, onto whom they had projected their own secret and organized practices of leadership. The feeling of suddenly standing there stark naked in the emperor's new clothes, so to speak, was subsequently drained of its reality by the fact that the 'noisemakers' of 17 June—as one of the functionaries we interviewed reported—had held out their hands to him after only a few days, in a gesture of friendly reconciliation. They proffered the illusion of a new normality in which no new problems surfaced after a solution had been found for the immediate economic grievances. The feelings of anxiety and rage among SED functionaires were transformed into the long-term conception of combining economic consumerism with authoritarian control.

How did they arrive at this mistaken conclusion that tends to corroborate the thesis that there could be no democratic alternatives under an SED regime? One clear and unmistakable factor is the presence of Soviet tanks, whose mere appearance on the scene put an end to the outbreak of feeling—but that does not explain everything, and is in particular of little help in trying to understand the upsurge of emotion. In this connection, I would like to return to those interviewees who admitted to having participated in demonstrations. All I noted earlier about them was that they had been shaped biographically, directly or indirectly, by fascism. This uniformity was probably an accidental element of our survey, yet it hints at certain determinant preconditioning factors. However, their reports do not give the slightest suggestion that they somehow formed an organized, intentional, or even conspiratorial grouping on 17 June; rather, they all joined already-formed and active demonstrations, and had left the isolation of their own hidden or repressed feelings to merge with the masses of people who had tossed off their bureaucratic shackles. This was a mass in which there was little distance between protest against a class-specific form of discrimination, the feeling of regained freedom and power of action, and the symbol of national unity. It appeared that by reunification it would be possible to overcome the totalitarian distorted twisting of the workers' movement at the hands of Stalinism, the one-sided placing of the burdening of reparations on the East Germans, and the necessity for a double rearmament in Germany. For those who professed support for the uprising, being swept up in a mass in revolt and the sensual experience of national symbolism constituted a feeling that they were being liberated from the suppression of their unexamined past.

We walked off the job, at the gate there were more and more of us, and by the time we got to Bitterfeld, we formed an impressive demonstration. Then all of us—many had got there before us, they'd come from all the different plants and factories, everybody was out marching—we all went over to the jail, the local court, they pulled the people out of their cells, set them free. The Gestapo—no, not Gestapo, the state security—they had their headquarters over near the train station, and the crowd got some people out of there too . . . Bitterfeld had never seen a rally like that one. There was an incredible amount of enthusiasm. The old national anthem was sung, people were deeply moved. Today too, when I hear that song, the *Deutschlandlied*, I kind of choke up. Even on TV.

I ask what was said there or what else happened, and he replies: 'We were all so preoccupied with ourselves, so happy and excited, that actually we didn't have any big thoughts about anything.'

Another interviewee, who said that he'd 'joined' the 'masses of people, the procession' streaming out of the factories, offered an answer to the same question: 'Well, freedom of speech, freedom of the press. . . . [pauses to think] . . . There were also some financial matters, connected with the local tax office. Let's see, uh, what was that? Oh, yeah, there was a demand for better pay, something like that.' But that was apparently not his problem—at least nothing that might have burned itself into his memory. Then he went on:

I was there at the square listening to what was being said. And the people were incredibly, incredibly enthusiastic. A high level of participation. So people were standing at the windows, and there were tears in their eyes. Then some [of the demonstrators] wanted to go in to negotiate with the local council, a group that had formed as a kind of head, and so the crowd dispersed. I went home, and that evening the Russian tanks came through town and occupied the factories. Yeah, and you clenched your fist in your pocket . . . Right, we felt something like the Czechs had when the German troops marched into Prague in 1939, in March. Outrage among the people, and everyone had to keep his silence. After all, the tanks were too much the representatives of the other point of view.

Basically, what they recall is movement, motion. This initially meant a movement against the rigid structures of a system that already bore some of the features of an anti-fascist educational dictatorship. But even more, what was meant was their own movement, their own emotion upon hearing the strains of the *Deutschlandlied.* After that, they didn't feel like developing strategies to pit against the regime or engaging in a struggle against the Soviets; rather, they went on back home, moved by their emotion. Or they recalled their duty and returned to the plant—in order to finish some task for society that they had left uncompleted. Or they called to mind, as someone who had been in the thick of the war expressed it, what the diameter of a tank cannon was and the size of the holes that its shells made. And in their pockets they clenched their fists. Another alternative—and this was the fifth reaction among our five informants—was that they developed a hatred for the Americans and West Germans, who they felt had abandoned them. None of them could forget this day of the public liberation of their feelings. Nonetheless, they all made their little arrangements with the regime afterwards, if only outwardly, following their experience of its actual power and the impotence of its opponents. Despite this, in the niches of their memory, 17 June was preserved as the basic experience of another potentiality—even though it had proved itself to be an impossibility.

The SED functionaires, whose total and complete surprise and disappointment, projections of anxiety, and impossible forms of 'proving their worth' and loyalty were sketched above, had no real support to fall back on in the official memory of the GDR in the face of the fragmented memory of the people and the arrested feelings of their opponents. This was not so much because official memory had canonized a legend that had little basis in reality, but rather mainly due to the fact that they themselves did not fit in at all with that legend. Roughly speaking, their life histories belonged to two historically specific categories of German contemporary history which the public self-image of the supposed 'leading force' of the working class had not anticipated.

Some had been functionaries in one or the other of the two factions of the old labour movement that had committed themselves in the form of the SED to the building of a socialist order. On 17 June, they had what was their 'Kronstadt experience', reacting with fantasies of power and constructs of conspiracy against a base that was unwilling to accommodate to their great life project. But it was far more than that. Our investigations in the field indicate that, in most cases, they themselves had been broken in biographical terms by the attempt to realize this project, and yet nonetheless had stuck to it doggedly. There was not a single one who had not been stripped of his office at least once during Stalinism; not one, whose anti-fascism had not been doubted—which had often led to a long incarceration or emigration; not one who, as a Social Democrat, had not been demoted or humiliated in another way by his party. For others, their historical confession of faith had at best the persuasive power of a rationalization.

The proponents of the SED legend who stated that the workers were not yet properly educated or who regarded a strike by workers as something 'essentially alien' belonged to that younger generation of SED functionaries that controlled the second half of the history of the GDR. This was a generation which had received its socialization in the matrix of Hitler Youth and BDM, and had undergone a subsequent moral-programmatic reorientation in the Free German Youth, though retaining the basic contours of their activist–executive behavioural type. This class of post-war party careerists had no direct familiarity with a living and lived tradition of the workers' movement. In their picture of the world, consisting of loyalty to the leadership, resoluteness in struggle, and moral black–white dichotomies (between brown and red), a grass-roots protest appeared to endanger the basic conditions of their conformity. It also seemed to tempt their past, often only just overcome after difficult crises. In other words, as a result of the totalitarian continuity, freedom to them seemed to be a form of fascism, and in their eyes the outbreak of feelings repressed after fascism was distorted into the threatening image of an organized counter-revolution.

Time and the preshaped specific angle of vision eight years after the end of the Third Reich, the national line of rupture running even through individual experience, and the dissolving of perspectives are all concentrated in two narratives related to us during interviews. They are not stories that can be easily accommodated to the formation of experience-types, but rather cast specific light on the prevailing situation—slanted, as it were, and thus revealing from the standpoint of marginal experience. One narrative sketches a picture from the angle of inner distance; the other from external proximity.

The first story is that of a Polish Jew who had fled from Auschwitz; he fought with the partisans and then fell in love with a German woman in Silesia in 1945. Since, with a small child, she did not want to emigrate either to Israel or the USA, they remained in the GDR, where he, independent and hard-working, became a successful and indeed indispensable entrepreneur. He kept his distance from

German politics, regarding the SED as authoritarian and corrupt. Nonetheless, he recognized the protection accorded to the victims of fascism by the authoritarian conditions—protection from a people whose relation with fascism remained a question mark.

There was a lot going on here. I can remember that the pavilion of the National Front that was set up in the square was on fire, there was lots of unrest. All the party badges were lying out in the street, flags had been ripped down, and all those political banners. I can remember that as if it were yesterday. But, well, I was so preoccupied back then with my own problems that, first of all, I didn't want to get involved in the thing. And second, I also noticed that various elements were waiting for something to happen—and not what I approved of. I want to emphasize that. Their wish was for a return of what had been before: the Third Reich, National Socialism. Nazi elements like that resurfaced here—

Did you experience that personally?

Yes, I did, personally.

Can you be more specific . . .

I didn't say 'everyone,' I said 'certain elements' that were part of it. In concrete terms, they were shouting 'down with socialism!', and I saw myself how they started to beat up people who were wearing party badges. I witnessed that with my own eyes, and people should admit that it happened.

Yes, but when someone shouts 'down with socialism!', then I don't consider that to be convincing proof that he's for fascism.

Look, when you beat people up—I mean, it doesn't always start right away with gas chambers. But it begins with stuff like that. Until they're in power, and then you can have a repetition of what happened. We also know that a number of people were killed. But I didn't witness that myself. Maybe party members who were too rigorous in their methods and who made innocent people suffer here too, to promote their own advantage. The way it used to be. And maybe it still is a practice even today. Those were facts that were not to my taste, and not in keeping with my own political convictions. I didn't like the political line back then, let me admit that in all honesty. But not that there should perhaps be a repetition of what had been. It always starts with something small, and then gets rolling.

There's a reason for that old saying: 'nip things in the bud'.

The second narrative is about two brothers, and was told to me by a former SED functionary. The story involved two very different sons from a Nazi working-class family in the chemical manufacturing district in East Germany. They were twins nonetheless, and had been given nicknames as kids: the 'roving wolf' and the 'little professor'. Both were in the Hitler Youth, served in the war and had been POWs, and both had returned afterwards to the industrial area in the East.

The roving wolf utilized the available opportunities, and the little professor was ambitious—he believed in the new politics and values just as he had believed in the old system right down to the end of the war. The roving wolf led the life of a rover: he worked on construction sites in the early years of the GDR, had various affairs, and was a good-for-nothing. The little professor proved his value in the lab and in the Free German Youth; in 1950, he started studying for a diploma as master technical instructor in chemistry while teaching the short course of the Soviet Communist Party in the party school at the plant. The date of 17 June found the roving wolf active in the local strike leadership; and the little professor making his retreat from the party school, preoccupied with the question of whether or not to remove his party insignia. Afterwards, the wolf went into hiding with relatives in Thuringia, and his twin brother advanced to the post of master instructor.

In 1954, problems arise. The wolf has long since reappeared, and initially everything seems to have been forgotten in the general thaw. Suddenly, however, the professor receives some information that the authorities are out to get his brother. He warns him, and the wolf heads West. A little later on, and with no apparent connection, the professor becomes involved in an intrigue with one of his superiors in the lab; he is stripped of his party membership and job, and forced, literally on his knees, to work laying tiles on a construction site. In 1956, this is declared to have been an oversight, and he gets back both his party card and his old position. But he feels he is stagnating in his job; an attempt to break out of the routine by writing wins him a prize but no success. In the meantime, the wolf has been working in the western branch of IG Farben and advances to the post of section head in a lab. The little professor, an ailing man, although still only in his fifties, is declared disabled, and retires. Afterwards, in the 1980s, he is allowed for the first time to travel to the West, and is happy that his brother, despite their separate paths and thirty years of silence, welcomes him with open arms. After returning home, he sits in his small vegetable garden and compares his tiny arbour with his brother's bungalow; his bicycle with the wolf's Mercedes; his pension as a disabled retiree with the pension and company supplement that awaits the wolf upon retirement, and which is ten times as much. In 1987, when I met him, he was still agonizing over the question of whether he should resign from the party. He is still active in social matters now as the chairman of a 'garden section'. However, since no one does anything there except tending their own little plot of land, what this means is that he himself has to mow the public lawns and remove the weeds from the public paths. At the meetings of the section, he gives political lectures; yet, attendance there has dropped to a third of what it used to be. He is outraged over the deterioration in discipline; worried about his next trip to the West; and hopes that Gorbachev will bring both an opening and a clamp-down. He sums up: 'There's nothing so disagreeable to a German as democracy!'

Based on our interview data, I have attempted here to attribute the opacity of 17 June in the memory of the GDR to a conflict of feelings. I have suggested several dimensions and contours for a possible typology; these lie at odds with the principal political interpretations and, in their historical specificity, were submerged and buried by those interpretations. Such conflict-ridden feelings moved only a small proportion of those who experienced 17 June, yet dominated the fragmentation and repression of its memory throughout the long decades of GDR history; repeatedly, the perception of the problem of democracy in socialism was viewed in terms of specific post-war constellations of German history as a result of this complex of conflicting feelings. In contrast, the vast majority of those who lived through 17 June put their experience at a distance, relegating it to the niche of uninvolved knowledge, so long as the threat of force permitted no other options and the new normality of socialist economy awakened expectations for gradual betterment.

In my view, it is highly doubtful whether this fragmented memory was able to constitute any kind of motivating background for the demonstration culture that developed in the GDR in the autumn of 1989, because that culture of protest was initiated and carried out largely by young people who had no direct memory of the primal conflict in the GDR. In a scenario quite different from that of the explosiveness of 17 June, it was only after many weeks, and in the face of the bankruptcy of the leadership in offering viable economic alternatives, as well as in the face of the opposition, that it was transformed from a push for democratization of the GDR into a thrust for national and economic unity. Of course, the gradual involvement of a portion of the older generation was of a certain importance for this shift in direction. Only in this regard can one speculate about the possible impact of the fragmented memory of the 1953 primal conflict on events in 1989. But the national plebiscite then became the expression of the degree to which current circumstances are bereft of any future. Yet, 17 June was stamped by the conflux of different pasts. It bore the joint imprint of an old society without any future and a new society—under construction—without any traditions.

Note

1. We—Dr Alexander von Plato, Dr Dorothee Wierling, and myself—were the first West German historians to get official permission in 1987 to conduct an oral history project on working-class experience in the GDR. Even East German historians had not been granted such a permission because oral history was regarded by the party hierarchy as a subversive instrument to arouse the retarded mentality of the people against the conscience of the established avant-garde. We wanted to conduct life-history interviews with some 25 workers and concentrated our research on 3 industrial centres: the old machine-building and textile city of Chemnitz in the foothills of the Erz Mountains in the south of Saxony, then called Karlmarxstadt; the chemical and coal-mining town of Bitterfeld in the vicinity of Leipzig and Halle, where most of the newer chemical industries of the inter-war period (then IG Farben) were situated; and finally Eisenhüttenstadt near Frankfurt an der Oder on the Polish border, a new steel city founded in 1950 and one of the prominent projects of the early GDR in the 1950s, then named Stalinstadt. But for various reasons we found it very difficult to define a worker (because almost everybody seemed to have been one at some time), and because we were less restricted than we had expected when we arrived in the field, we finished up with 150 interviews with people of both sexes born between 1895 and 1935, the minority of whom had been workers throughout their lives, but the large majority of whom had come from working-class origins. Some 40% were members or functionaries of the SED—this is double the proportion of party members within the adult population of the GDR—and another 5% were members of the other four parties of the ruling bloc. We usually found interviewees initially through party and trade-union networks (roughly 40%), and then through churches, personal acquaintances, and (for more than 40%) through presenting our project at old people's lunch clubs. At one-sixth of the interviews an East German historian was present to check on the project, but in practice proved very liberal and interested in our method. Interviews averaged 3 hours, up to a maximum of 10 hours. They began with the informant's open life-story narrative, with few questions, but for later stages we had a long checklist of family, work, social, and political questions, some especially designed to encourage reminiscences of particular political events during the times of National–Socialist and Communist rule. The export of the interview material was exempted from frontier inspections. Of course, the very special circumstances of these interviews make them a singular historical source; such interviews could not be repeated or continued in the same way after the collapse of Communist rule in East Germany; since 1990 there have been, however, follow-up interviews with some of our previous interviewees, the majority of whom did not want to be reinterviewed under the new circumstances. Even though people might now recollect their experience more freely than before, the impact of collective on individual memory is just as important now as before, but different, and most people were reluctant to show us both sides.

4

A German Generation of Reconstruction

The Children of the Weimar Republic in the GDR

DOROTHEE WIERLING

My interest in the following text is in probing some experiences of Germans in and under National Socialism, as well as in the immediate post-war period. The focus is on the 1940s, a decade in which their personal and political outlook underwent a radical change. How did Germans try to explain this sudden change to themselves; and what conclusions did they draw from this as regards their attitudes toward German post-war society? How was this society shaped by these factors? And what perspective does the war and post-war generation have on the swift changes now taking place in Germany?

The problem of the biographical continuities of individuals and of the German population as a whole through the ruptures of historical experiences has been broached by historians and sociologists of West German society in a number of different projects. There has not been any corresponding study for the GDR, until recently—nor is this a surprise, given the problems regarding access to source materials confronting even traditional research on recent history there, up until 1989.

In January 1987, Lutz Niethammer, Alexander von Plato, and I received permission to carry out a field study on life history with 150 male and female respondents in the industrial province of the GDR. The persons interviewed were primarily workers, housewives, and white-collar employees.[1] From this corpus, I have selected two life histories in order to explore questions regarding continuity and rupture in the experience of the so-called *Aufbaugeneration* (generation of reconstruction) in the GDR.

The respondents were born in the 1920s; their earliest memories go back to the Weimar Republic with its hopes and crises, while their conscious political socialization is closely linked to National Socialism. In 1945, this generation was young enough to take up the challenge of a new beginning in post-war Germany, but at the same time it was extremely hurt by the experiences of war and defeat, and more sensitive and vulnerable to the possibility of a new political involvement than their beaten parents. It was this generation that not only immediately and very practically reconstructed the country in the Soviet Zone, but also helped to establish the economic and ideological system of socialism in the GDR from the 1950s and with a certain success in the 1960s. They were the main social basis for the masses of new cadres in the parties, in industry, co-operatives, and the town administrations, until in the 1970s a younger generation with exclusively post-war memories started to take over.

Johanna Maczek was born in April 1923 in Eidlitz, a small town in the German-speaking part of Czechoslovakia. She retired in 1985 and lives alone in a three-room apartment in Wolfen, not far from her former workplace. The interview begins with her request for the interviewer to determine the framework of her narrative.

You have to tell me what you want.

No.

Should I tell it chronologically or?

Just start telling your story, Ms Maczek, just begin talking now.

But you have to tell me, OK?

Just start telling the story, simply tell it the way you . . .

Because we brought it up, well, whenever I think of my youth and the circumstances back home then, there were never any problems, though there were so many religions and philosophies. Or, uh, people were mainly Roman-Catholic, if I think of the schoolclasses, there were kids, who had no religion, see, there were very few, uh, Protestants, and then there were some Jews. Yet that was never a topic for conversation, and was never a problem.

When the interviewer refuses to give her guidelines for narration, Ms Maczek surprisingly begins her remarks about the different religions and philosophies in her youth with 'because' (*weil*). She seems to be uncertain for two reasons: first of all, the situation is an unaccustomed one, namely to talk with a younger woman from the West—one whose 'philosophy' she doesn't know. She links up this sense of current insecurity with her experience as a child in the 'border area' (*Grenzland*), as she refers to that area once at the end of the interview. What was involved there was not only differing philosophies in the sense of religions, but rather national differences—indeed, racism. Thus, Ms Maczek refers right at the outset to the unresolved knot in her story—that certain 'something' which she regarded as unproblematical, and which was never a 'topic for conversation', yet led to the decisive break in her family and was the central topic of her life history. The period from winter 1944–5 to August 1946 is the 'terrible time' that she had to endure; the 'worst thing' during that span was when they were compelled to leave their house, in May 1946. She had grown up as an only child in that house, and her parents had offered her 'the maximum'. She spent a 'completely untroubled' childhood. Her grandparents lived right nearby: her mother's father, a Social Democrat, who made sure early on that his children would better themselves, and her father's mother, who was the pride of the family, an excellent wet-nurse.

So Johanna grew up in a social environment of love and liberal understanding, oriented toward social advancement. She recalls the year 1938 as one in which the tensions between Germans and Czechs were deliberately and artificially fanned from outside. The latter tried to reduce the German majority by means of settlement, and the Germans attempted to distinguish themselves from the Czechs by means of dress. The seemingly senseless conflict led, however, to the annexation (*Anschluß*) which was welcomed by the family, Johanna's father having been a member of the NSDAP (National Socialistische Deutsche Arbeiter Partei) since 1935. Soon after, Johanna took her final exams at the commercial school she was attending and got a position in a bank she had admired for many years because of its festive, neo-Gothic arches. Though the National Labour Corps was an unpleasant interruption, Johanna later profited from the war in that she, like other women, was able to take over the posts of men who were at the front. 'They didn't think the girls could do it, but it went quite smoothly.' However,

> that didn't last very long, 'cause, er, in 1945 what happened was that, er, practically right after the collapse, the first women were dismissed, and that was simply . . . That was the terrible thing, that you experienced and saw it all. And that mothers simply had to leave [their homes] within twenty minutes, and then were standing there on the streets with their kids. There were such heart-rending scenes . . . I think I'll never forget that, as long as I live.

If before, people had worn white stockings up to their knees to 'look German', now white armbands became the badge marking the discriminated. The newly-appointed mayor came to her father, who had been a block captain and a representative of the Nazi mayor, and said: 'Come on down for a second to the city hall.' They never saw him again and it was not until years after, that the family finally learned he had died half a year later. Johanna received a certificate giving her permission to continue working at the bank. The 22-year-old now became the breadwinner, protector, and authority in the family. 'Starting in 1946, I was actually the one who determined what we did. Her father had tenderly called her dainty mother and his powerfully-built daughter 'my little girl and my big girl'. It was now the big girl who determined the time of departure, when she 'could no longer stand it there'. She had put away silver coins in advance, but they were not permitted to take these. She was the one who arranged for seats on the train, for herself, her mother, and her four grandparents.

Johanna and her resolute wet-nurse grandmother kept their heads about them, while her paternal grandfather, who was 'a good-for-nothing' in any case, finally went out of his mind while they were fleeing. He had to be stopped several times from jumping off the moving train, and in Germany he ran away from the hospital where they had brought him. He was found dead, lying alongside the railroad tracks—where he died, Ms Maczek believes, of weakness and despondency.

Her maternal grandfather also became 'childish': he would roam through the new, unfamiliar neighbourhood, forget the time, get lost in the dark, and wake his daughter, Johanna's mother, who was sleeping after coming home from the night-shift, so that she would make him something to eat. Johanna tried to 'gain a foothold and settle down' with her family, still suffering from the shock of 'resettlement' (in the interview, we were not able to agree on the term to be used: she rejected the words 'flight' and 'expulsion').

This also meant that her own plans and designs for living, which appeared shattered by the terrible collapse of 1944–6, could not simply be resumed. Two months after her first numbing experiences as an unskilled worker, she met someone in a stenography course at the evening adult-education college. He was from the cadre section of the plant she later worked at. She became a stenotypist and bookkeeper there, and then fought her way up the ladder. In 1953, she advanced to the position of special clerk; and to section head in 1960. In 1964, she finished a course of studies in economics and later, in 1968, became department head. To her surprise, her advancement was successful despite her 'deficiency'—that is, the fact that she was not a member of the socialist SED party.

Significantly though, this deficiency was based on a 'resolution' that she had made 'after the experiences I went through that one year', when her father, whom she describes as friendly, harmless, and popular in the neighbourhood, died so unnecessarily, guiltlessly, and accidentally, while others who had been arrested, had a much easier fate. Many were released after a short time or were not even placed in camps, but rather left somewhere at the side of the road by negligent soldiers. The unpredictable quality of events, and their opacity, which led to the death of her father and the loss of her home, rendered politics something despicable and disgusting. Johanna Maczek decided she must avoid any contact with them, even should that mean fewer chances for advancement, or the need for her to make a greater effort to get ahead. But what made her seek a career anyhow?

When she talks about the central event in her life history, the segment 1944–6, she thematizes her own loss by describing that of the older generation:

> But, as I said, the older folks had a much worse situation, 'cause of that, since they had built everything for themselves, right? They worked their entire lives for that. Hm, and actually, they worked so hard just so that the children would have things better some day. So anyhow, the upshot was that I was poorer than anything that had ever been before, right?

Thus, her own success and advancement became a kind of obligation *vis-à-vis* the family—a guilt toward earlier generations, that must be redeemed. One wonders whether, right from the beginning, there was also proof, alongside this familial logic, that Ms Maczek believed she was indebted to the new society. Again I look for this answer in that 'terrible period' of 1944–6, with which—even today—she has not completely come to terms. I make a reference to the connection between her expulsion and the criminal war of the Nazis, but she responds with reserve. She certainly regrets her youthful thoughtlessness and the ignorance with which she lived through the Nazi period (50 kilometres from Theresienstadt). She also is willing to accept the connection I suggest, but only as something abstract, not as an element she feels is part of her own personal distress. She characterizes the connection as something whereby they suffered the consequences of 'what others did'; they paid the price.

> *When did you hear about the war crimes of the Nazis?*
>
> Actually, it is not until here. Only here.
>
> *Can you maybe recall what you heard back then, what the first news was? What impact it had?*

Yes, that a lot of people, a great many, died and were put to death, right? And er . . .

Did you believe that, immediately after you heard about it?

Did I believe it? Well sure!

Yes?

Yeah, because I mean, all that other stuff, that was sort of the price you pay for this, a kind of bill, right? Practically speaking, that, er that was a consequence of all those things, it, I mean, everything that had happened sort of added up to this. Well, and that people who live somewhere in a border area, that they had to suffer, that's happened before, I mean, or Silesia, see, like, I mean, that happened to them many times, didn't it? But we didn't think [laughs], didn't think it was possible.

Yes.

But it's really so: you can only absorb in an intensive way up to a certain point, it bothers you and oppresses you and depresses you and, and from a certain point on, I don't know, it somehow just dissipates, it vanishes. You somehow just shove it away from you, I don't know.

Johanna Maczek was able to believe the reports about the crimes, indeed had to believe them, because only by doing so was it possible to explain why expulsion—which she had never thought possible—had actually taken place. Since she was presented with a real 'bill' for a 'sum' of crimes, those crimes must be just as real. But Ms Maczek cannot interpret this in a sense of punishment for guilt incurred. Rather, it was her family's turn—because they were 'there', in the 'border area', where no distinction is made between the innocent and the guilty. But while all this bothers, oppresses, and depresses her, she decides to live and to show through her life that she has a right to professional success, social recognition, and respect. As though, in this way, a part of that bill was cancelled—and thus a portion as well of what it was meant to cover.

SED party secretary Rudolf Kamp, born in 1929 as the son of a baker in a small town on the then German–Polish border, was interviewed twice: first by Niethammer and von Plato, in the presence of one of our colleagues from the East Berlin Academy of Sciences; and later by me, focusing more on his private life, his childhood, and youth. In the second interview he opened up more, talking to a younger woman who did not seem to be all that interested in politics or too critical of the GDR system. After the first interview, three hours in length, all those involved had taken leave of each other, politely, exhausted. It had not proved possible to break through the barriers of caution and mistrust, erected by Kamp. He had largely made use of fixed formulas of expression to shield his life, belief, and party from the sceptical questions of his interlocutors, historians from the West. The colleague from the GDR was also disappointed: the interview had yielded little more, he felt, than a few personal comments on the 'party apprenticeship year'.

Since the 1950s, 'the party' had shaped and determined Kamp's life. On its instructions, he had exceeded the norms in the potash mine. He had agreed to be assigned to the Police Guard, taking a considerable cut in salary. From there he was delegated to the steel industry, studied engineering under heavy strain, and accepted party functions over a span of thirty years, until he fell ill—psychosomatically—a number of times. Why did he always speak about the party, when asked about his private life?

Yes because that's my life.

This is your life?

That's my life. I told you: when I decide for something, then it's 100 per cent with me . . .

His course of life was 'shaped by the party' and he is unable to identify any contradictions with his personal wishes and desires.

So I never had to think about things. They always told me: 'First you go there, then you go there.' (laughs) 'Go there, then go there.' So I'm, in my case, the party has shaped me. And the party's also helped to determine my life somewhat.

This is not at all unsettling for Kamp—on the contrary. He tries to explain to the somewhat confused interviewers how it is that he has placed himself so completely and unconditionally at the disposal of the party, in order for them to understand why all questions about 'problems' miss their mark, and why party and life are for him one and the same.

For Kamp, the party is the body, the agency bringing his life history to harmony with world history, in that he places himself under its laws—which are oriented to the laws of social development. This is not governed by arbitrariness, accident, or forces of nature, but rather by science, regularity, and clarity: in a double sense, by law. In contrast to the menace of nature, at whose mercy he feels himself powerless—'a flash of lightning, whether or not I want it'—he imagines society to be a structure whose regularities and processes can be managed by human action—by humans helping to implement these regularities.

For his own person, this means accommodation, adapting oneself, conforming. For his political work it means 'education' (*Erziehung*)—just as 'adapting oneself' means self-education or re-education, especially in connection with a previous chapter in his life, namely 'education and moulding in the false, opposite direction', which he wished to overcome in the 1950s by dedicating himself '100 per cent' to the new, correct direction. From that point on, society in the GDR became a playing field on which he controlled the rules of the game and helped them to dominate. To the Westerner listening to him, these rules may appear to be boring, confusing, or unrealistic. Yet, Kamp understands something about this and presents them indefatigably; defends and explicates them. Lovingly, he unfolds the system of 'socialist competition'—with its code numbers, funds, premiums, and commissions—which he organized for many years for the factory combine, spreading them out before his audience of uncomprehending Westerners for over half an hour. The democratic character of this system appears given to him, established by dint of its fixed, predictable, hierarchical structure alone. 'This simply is our principle of voluntary choice, our principle of conviction, that they [the workers] know exactly what codes they have to fulfil.' And Kamp also loves the notion of the economic plan and the process of 'planning discussion': when the final plan has been 'announced', 'then that's planned. That's what planned economy is. And that's good.

This is how Kamp allays his own fears. Because, shy and amicable as he really is, he finds himself constantly in the position of a person who has to implement the 'plan', the policy, the law—and against the natural needs of the living, breathing workers. He talks in a distanced way about 'our people' (*unsere Menschen*), whom he would like to 'confront in an open manner'. 'You have to try to find the right sort of contact with them', which is not easy if you are not interested in soccer and are confronted with their aggressiveness when their favourite team has lost the game. Kamp is an affable guy, because he is afraid of all those people he faces as a party hack. When in the early 1950s he exceeded the norms in the potash mine under party orders, the other men showed their displeasure by spitting in front of him and dubbing him 'a norm breaker'. But today his job of education and schooling is still difficult, because

when they come and you show them these laws [of historical development, *Gesetzmäßigkeiten*] and tell them, OK, that's how it's going to be, that's how it can be and must be. Not just that we understand it that way, but rather: it's simply gonna be like that, because that's the law of development, can't be different. Then they always say, 'Man, you talk this jive. Just look outside, see the way it is! And the way you're telling it here, hey buddy, that ain't . . .' Then you've gotta, well, you have to try a little too, I mean, because in reality, in practice, it still looks a bit different from what Marxism–Leninism teaches. But that's life.

Kamp's life in the GDR was shaped by the problem that the laws he believed in could only partially be reconciled with the reality that the subjects treated in the party courses referred to. But still he believed he depended on the impeding scaffolding of Marxist–Leninist 'regularities'. Why is this?

It was my intention in the second interview, to lure Kamp from his inflexible defensive position. The conversation dealt mainly with his childhood and youth, that is the period of National Socialism, the war, and the immediate post-war years. The end of the war and the flight of the family play a key role here.

In narrating these events, he comes to the story of his private capitulation as a soldier. He had joined the trek of refugees along with his father (his mother and sisters had left earlier, while he stayed behind to bake bread for the units building the trench fortifications). When they reached the Oder River, Kamp was taken aside once more by the army, 'selected out', and sent to a special military training camp (*Wehrertüchtigungslager*) where he was put through drills and trained for close cambat; he was then shipped out to the front on 6 May 1945 in order to 'defend Prague'.

Two days later, his unit was disbanded, and Kamp, now in civilian clothes, made his way with six other buddies to Wurzen near Leipzig, where his family was. In his first interview, he had already hinted that during this retreat in Czechoslovakia he had had 'a lot of difficulties' and got 'knocked around a bit'. I asked him to give an example. He related the following.

While they were looking for places with aid and food for refugees, the boys came to a school

and there were Czechs in there. The school had been occupied by Czechs, there were Czechs in there and we were probably just what they were looking for. OK, first they received us, you know, and then, er, and, like they started up: 'You German swine, we're gonna take care of you characters'. So we were given spades, then they took us out into the fields, and we had to dig a big hole there, a large pit. Then we had to go around the field and collect stones. After that we had to place them out in front of us. So we thought, OK, kids, this is it, we're had it. Yeah, and then we had to take off our shirts, strip, naked from the waist up. After that, they brought over a lot of big pictures of the Führer. But they would smack us, hit us, wallop us, whenever we weren't digging fast enough, giving us a thrashing all the time. Then this guy came over, he was carrying a bull-whip. And they propped up those portraits of Hitler, ten, maybe 15 metres long, in a row. And that guy, that guy ordered us to throw stones at the pictures and we were supposed to shout: 'We owe this all to you, you swine!' So naturally we had to do that, fling stones at the portraits of the Führer, and up on top was the guy with the bull-whip. So they walloped us and then we were allowed to get dressed again . . . The next day we were overrun by the front, the front just caught up with us, and we went back home with them.

He arrived a few days later, on 18 May. Two days after that, on the 20th, he celebrated his sixteenth birthday.

Kamp had been meted out a terrible punishment: the fear of death, together with debasement, humiliation, and the compulsion to desecrate the previously so worshipped idol of Hitler, before the eyes and under the command of those who earlier had been subjugated. All this must have deeply disturbed and hurt Rudolf Kamp. His encounter with the Red Army must have seemed a kind of rescue to him, a sort of protection, compared with the unpredictable, individual violence of the Czechs. The Soviets asked if they had any watches or bicycles (which they did not), and then sent the group on its way home with a properly-stamped transit permit. Their way led them through Czech towns and villages, but the Russian guards prevented any acts of revenge from being carried out.

Seen against the background of this narrative, it becomes understandable why Kamp had already mentioned in the first meeting that he was happy that 'the friends[2] had caught up with us, and that we'd gotten a bit of backing, support from the friends'. Thus, Kamp also noted, that the 'rethinking process' in 1945 had begun with the first 'contacts with the friends, and also with the treatment back then'. His treatment in Czechoslovakia had shown him, 'what a terrible reputation one has in the world, as a German'. For that reason, he was astonished that the 'friends' had been much more generous, despite all the havoc and damage the Germans had wrought upon the Soviet Union. And he regards this as the result of

the fact that the process of education was further along there, and I read a lot of books then, like *Spring at the Oder River*, very beautifully written. As I said, there was that process of rethinking in the Soviet army, implemented by the Communist Party—. . . that you shouldn't—demand an eye for an eye. So these books helped me a great deal in my own process of rethinking and discovery of a class standpoint.

Kamp views the lack of a sense of revenge in the Red Army as a product of a process of rethinking, one that he attributes to the education by the Communist Party. Apart from the experience of his personal surrender, he does not allude here to any other experiences of his own, or to reality—but rather to books that awakened in him ideas about the process of education among the Soviets. But he had his own real 'Spring at the Oder River', when he crossed it to find his family in Wurzen, where the Russian army had just arrived.

So there were rapes in Wurzen?

Sure. . . . but not among relatives in Wurzen itself, not that. The first measures [of the officers against that] that I witnessed, they were very hard, very hard. Among their own people. So that . . .

What sort of measures?

They dealt very hard with their own men when they got wind of stuff, like when there was some raping, guys getting drunk, then going out to rape. Well, when they heard about stuff like that, I mean the Secret Police, then they were tossed like cattle [on a truck], then they didn't spare their own party either. Very rough, yeah, very rough (*hart*).

Asked whether such measures had not relieved and frightened him at the same time, Kamp answers in the affirmative. But later on he talks again exclusively about the horror of the wild violence of the individual and the attempts at education by the party. Kamp's process of rethinking did not simply engender in him a love for the Russian people—whom he describes, using the typical, innocuously racist stereotypes: 'friendly, good-natured, like their liquor, irascible'—rather, he views the Soviet army as the first new power that has come to re-establish law and order; a power that provides the first semblance of security once again for his endangered existence, giving his damaged identity a firm renewed hold.

He had also found this security as a 10-year-old in the process of his separation from the familial ties of childhood through his participation in the National Socialist children's organization (Jungvolk). He still recalls: he was accepted on 20 April 1939 (Hitler's birthday), while the formal oath-swearing ceremony was a few weeks later; the formula contained words like 'young, hard, silent, faithful, there was something in it like "honour" ', and then he got a camping knife 'on the blade there were the words "blood", er, "blood and honour" '. Training was 'very hard, very hard, I mean very hard. Service also was very hard, the cross-country field games were also very rough stuff.' Again and again, Kamp uses this word 'hard' (*hart*) that for him had already characterized the rigid politics of the Soviet officers after 1945. In the field games, what was at stake was the 'life ribbon', an armband or kerchief that was then kept by the victor. It stood for 'life or death'. Kamp was often scared; often needed courage. On the other hand, he had experiences of strength too, like when he marched through the streets of towns and villages with his unit and their band:

> that was really powerful, the window-panes were almost rattling in their frames. That was really powerful, very impressive. And then we had these cross-country field games on a grand scale, I mean a thousand or more men, kids. That was very, very powerful.

These adjectives, 'powerful' (*gewaltig*) and 'hard' (*hart*), describe the basic experiences in the authoritarian, strictly hierarchical organization of the *Jungvolk*, where masculinity was experienced and drilled in simultaneously with pain and joy, and the cross-country games were a matter of life and death. Kamp, who praised the comradeship of the leaders in the Jungvolk—he himself, despite frequent, long illnesses, made it to the rank of squad leader—did not experience the collapse of 'the powerful thing' until the war's very end. It was the collapse of the security and power that he had felt by belonging to an organization which was a protective collective, although, when he went through basic training for close-combat in the Wehrertüchtigungslager in 1945 and was at the mercy at night of the harrassments of the intoxicated camp commander, he felt alone. Finally, the only advice he was given, was to set out alone on the way home, right across the territory of the hostile Czechs. And they captured him and staged their own game of life and death.

Kamp did not regain that sense of security until 1949, when he joined the official Free German Youth (FDJ), and became a candidate for membership of the SED in 1952, finally, he entered the newly founded People's Army in 1955. He found it again in being part of an organization, one that demanded hardship and was rough on him, yet at the same time offered protection and the certainty of having a solid place in a hierarchical structure. He needed 'the powerful thing', and now he had it once again—in an authoritarian, highly regulated form.

Moreover, he had found something else: an explanation of the world, a certainty about the future of the sort that springs from closed ideology. Rudolf Kamp had joined the Jungvolk in 1939, a short time before the outbreak of the war. Like many boys and young men, he was

very enthusiastic about the war, because the entire education question as it was dealt with in the [Jungvolk] service also went in that direction. The battle actions were evaluated, there was a lot of commentary, it was explained, we even modelled mock attacks in the sandbox. So this proved to us why it could not be different in any way—why we simply had to win, be victorious, and, er, and . . .

Even today, he is still impressed by the effectiveness of this propaganda—about which he was repeatedly reminded at the time of the interview, because West German TV was broadcasting a number of old newsreels from the period. But just as the workers contradict him today when he tries to inculcate them with the laws according to which socialism will be victorious, the young '*Pimpf*' found an indirect contradiction as from 1942 he used to listen to the sceptical stories from the front, told by his sisters' husbands home from Russia on leave. He thought about it then, wondering,

maybe they don't see things correctly, or maybe they were too slow, they were NCOs (*Unteroffizier*)—I mean, what does an NCO know about the overall situation on the front and everything that's going on? That was what went through my mind back then. . . . He doesn't have the overall view of it all. Sure, he sees his small section of the front. Whatever unit, OK, they have a segment of the front, and when he says: 'sparsely occupied', how can he judge that, how can he say that the war is lost for us?

Just as Rudolf Kamp consoled himself during the war with the 'overview' of the situation, which he thought he had acquired as a result of his mock battlefield games in the sandbox, he relied—as a party secretary—on the notion that the overview of things, which he had gained through the organization, rendered him superior to the other men in the plant.

A lot is criticized by non-party people, for various motives, due to an inadequate overview of the situation, insufficient understanding, or a lack of proper information. They often criticize things that, when you know the true picture, look quite different. There are those situations in which you have to listen to criticism, very rough sometimes, very hard. They're very hard, very open . . .

Again Kamp uses the word '*hart*', now to characterize the attitude of the workers towards him. During his life in the GDR, Kamp never lost the notion of being threatened by social and political hardship (*Härte*). In 1945, as a 16-year-old, he was still too young to dare to find strength and identity outside a fixed stucture of authority; in 1989 he was probably too old to try.

The 1940s, as they are represented in the life histories of a man and a woman born in the 1920s, are marked by typical experiences of collapse, loss, and violence. The flight as refugees led to the loss of *Heimat* to giving up their native town, neighbourhood, house, and furniture. As a consequence, in certain cases families also broke up: individuals fell ill, became lost, died. In Eastern Germany, the collapse took place accompanied by considerable violence. Most 'expellees' from the East were initially there, and conquest was marked by individual acts of violence, especially against women. The severity of the intervention into daily life had a particularly powerful impact for yet another reason: the mighty occupiers were feared and at the same time despised as uncultured barbarians; they had long been the object of traditional racism. Politically, they represented the Bolshevik 'main enemy'.

The end of the war was received with apathy by many Germans—by some with numbness, as a kind of punishment; by others it was viewed merely as a blow of fate. But on the other hand, dismay in the wake of the collapse was strong, because along with the external collapse and havoc, ideology also lay in ruins. Faithful Nazis saw themselves vulnerable, stripped of their collective protection, without a Führer, without an explanation for the world. Especially for the younger generation, the collapse came as a tremendous shock, while the older and non-political had had a premonition of the impending doom for a long time. Indeed, the experience of the 1940s seems to have been determined most distinctly by generational affiliation. Those born in the 1920s were affected very differently by the debacle of the 1940s: it often hit them harder than their parents' generation. On the other hand, despite disturbing disorientation, they frequently develop the energy for subsequent acceleration of the activity with which they realize their life-plans and designs. While the very elderly tended to collapse personally, in the wake of societal collapse (examples like that of Johanna Maczek's grandparents are frequent, and often more dramatic), the middle generation, especially its women, are often bound to responsibilities towards their families and concentrate on the challenge of everyday life, thus avoiding psychological crisis and reflection on political responsibility.

The situation is different for the younger generation. The Germans born in the 1920s, like Ms Maczek and Mr Kamp, differ from those born earlier not only with regard to their experience of the 1940s, though this experience was preformed by their own lives through the Weimar Republic and first Nazi decades. They also develop a specific attitude towards the newly established post-war society and they are probably affected in a special way by the recent political changes. These generation-specific experiences will need more systematic consideration.

In historical memory, the Weimar Republic often appears as a society of almost permanent political and social crisis. But still, especially for the working class, it was also a phase of growing positive social perspectives—at least until the global depression of 1930. Better education and growing consumerism had their effects on the options workers saw for their children. Lower birth-rates, longer schooling, and job training for girls expressed these new perspectives and answered the need of modern capitalism for more and better-qualified white-collar workers. The growth of a new lower-middle class and a certain levelling-out of lifestyles of the urban masses led to what then was called the *embourgeoisement* of the working class. Those born in the 1920s felt this new attitude as an encouragement for them to try individual social advancement. Ms Maczek is a typical example of this experience.

For many Germans, National Socialism reinforced this confidence in the future. For young people, the offer of positive political experiences and identity in the Hitler Youth was without parallel: it was the attraction of a hierarchical organization that seemed to have overcome class borders, helped them to find a strong position against parents, and also satisfied the need for passionate devotion to a cause and a leader, by its quasi-religious rituals. In contrast to their parents, who had been defeated in 1918, they, the young, found images of power they could identify with and which stood for a glorious future. This is why the time spent in the Hitler Youth organization is very often remembered as a good time, for girls as well as for boys. Some stress the unpolitical sociability, others, like Kamp, admit the deep attraction of Nazi ideology and the powerful as well as brutal comradeship.

In 1945, this is abruptly followed by a sudden and shocking weakening. Germany, the Party, the Führer—all these projections of power collapsed and left a disturbing vacuum. Parents were likewise weak and helpless, if they were present at all. But at the same time, this situation was a starting-point for a new experience of strength: young kids often took over, not only during the flight from the East, like Johanna Maczek, but also in the family itself, where they often got into the position of sole breadwinner. It was perhaps this new social role, the need to earn a living and take over responsibility, that seemed to help mentally against the terrible loss and the lurking depression. In this way Johanna Maczek, who in 1945 was in her early twenties, took over the role of the parent generation, focusing on her job and their preparations for 'resettlement'. For, how to explain the past, how to interpret the present, and how to find a new perspective for the future? Ms Maczek later regained control over her own life by concentrating on her private reconstruction: working for a home and a career. Mr Kamp, however, after spending three years in an obvious vacuum (three years about which he has nothing to tell), found a substitute that provided him again with a social and ideological framework.

'Reconstruction': this was the magic formula for a large part of this generation, which sought a fixed place in the new society, after the decision to stay in the Soviet Zone had been made. This meant not only individual resettlement, but for many it included the intention to build up a new economic and political system, a clear alternative to National Socialism. The latter had proved to be on the losers' side; now Kamp tried to be with the winners of history. Others followed a leftist familial tradition, and the initial engagement was often enforced by pride and obstinacy against the West, where things seemed to be so much easier. Political involvement meant being caught in a close net of functions, activities, and schooling within the milieu of the avant-garde. Those who were less interested in politics, but more engaged in the economic and social development of the country, were also integrated into this milieu: the more so, the more they advanced in their career. Moreover such a career was offered to almost everybody in the 1960s, especially to women, who, like Ms Maczek, were encouraged to qualify for management positions, when the goal of a socialist economic wonder was the task of the day. So the generation born in the 1920s forms the basis of a first generation of socialist managers and cadres, most of them staying in the town, in a plant, a co-operative, that is 'at the front' as they put it in the interviews, expressing the daily need to confront an unwilling and mistrusting population.

The successful part of this generation seemed to be so thanks to the socialist post-war Germany. Although often this success had already been prepared for them in their pre-war childhood and youth, their social advancement after the war and in the difficult post-war period seemed to be a genuine socialist outcome—even the unpolitical among our interviewees stressed the opportunities they believed they would not have been offered in the West. They built a loyal and grateful body of core élites, with little initiative, cautious and silent.

The generation of reconstruction was not among the supporters of the political opposition in the GDR. We do not yet know how this social group experienced and reflected the political change that took place in 1989. They were hit by the collapse of their state, at the age of retiring. And although this means less social danger than unemployment for the younger East Germans, we can assume that the older ones are probably among those who gain little or nothing from the new situation. Their loyalty to the system now makes them outcasts, and their economic work of reconstruction is shattered.

Though the two life histories on the one hand present this type of generation of reconstruction, on the other hand they differ significantly. Under the shock of 1945, Johanna Maczek learnt to see politics as a fatal threat, and consequently refused to engage politically. But she was twenty-two years old, an adolescent, and the only breadwinner and head of the family. Her private strategy for success forced her to renounce partnership and motherhood. Instead, by pursuing the social logic and fulfilling the prophecy of her childhood—upward mobility—she substituted for her father, whom she had lost at the end of the war. Her political loyalty towards him, however, at the same time put a limit to her advancement.

Kamp, on the other hand, had learnt that only collective discipline and organization helped against the arbitrariness of life. As a loyal supporter of the system, he did not seek or gain social advancement (or at least only little) but the security that derives from total submission under a political purpose in power. He was sixteen years old when the war was over, the youngest in a so-called complete family with two grown-up sisters. There was no need for him to take over the burden of early responsibility so that he must have felt the vacuum that the past had left even more.

In spite of these differences, both life histories show the energy and the drive of a generation that had been frustrated and hurt by National Socialism. But their strategies and needs led them into another trap—their personal impasse was that of a society: the GDR of stagnation until collapse. And while during their post-war life under socialism their memories of National Socialism and their traumatic experiences during and immediately after the war had to be kept private, so now, it seems, there is little space in society for an open discourse about what the then new society of the GDR meant to this generation, and why it failed.

Notes

1. A first evaluation is Lutz Niethammer, Alexander von Plato and Dorothee Wierling, *Die volkseigene Erfahrung: Zur Archäologie des Lebens in der Industrieprovinz der DDR*, Berlin, 1991.
2. 'Friends' was the term used for the Russians, not only, but mainly, by SED members. The term derives from membership in the Society for German–Soviet Friendshp, one of the largest mass organizations in the GDR.

5

After Glasnost

Oral History in the Soviet Union

DARIA KHUBOVA, ANDREI IVANKIEV,
and TONIA SHAROVA

Remembering, whether personal or political, has its own special social and historical context in each country. The relatively untroubled, unpolitical British confidence in the worth of the past thus contrasts with the overtly contested claims to political legitimacy of post-war Italians, the enforced silence of Spain under Franco, or the ambivalence or deliberate silence of the wartime generation in France or Germany. Yet the Soviet Union is perhaps the most remarkable case of all: a society, probably unique in the whole world, where remembering has been dangerous at least since the 1920s. The cumulative effect of fear of public remembering, together with the fact that so many families had members who were politically oppressed, and so had bitter memories, is very difficult for Western historians to understand. It is not just the political impact—although recording memories in Russia certainly has had and still carries political implications—but also the dramatic long-term effect on personal remembering. For both reasons there has been no easy social tradition on which to build oral history. With the beginning of glasnost in 1986, it became politically possible for the first time to start openly with oral history. Earlier, there had been some clandestine attempts to collect memories, some of which were bravely published in the West through subterranean networks, and there have also been recordings done with Russian *émigrés* in the West.[1] But open recording and publishing in the Soviet Union itself has only now become possible.

This paper is based on the recording of a joint talk given at a meeting of the National Life-Story Collection at the National Sound Archive, London, on 21 Mar. 1990. It incorporates part of the discussions and some of the comments of Paul Thompson as chairman, and also some writing by Daria Khubova on the earlier history of the use of oral sources in Russia. She was founder of the Oral History Club and leader of the expeditions described.

While there are many individuals working from different institutional bases, two organized groups are especially significant. The first is the Oral History Club at the Moscow Institute of History and Archives—formed by young historians with the support of the institute's politically prominent radical director Yuri Afanasiev—which organized the first Soviet oral history conference in September 1989, and was able to open its own office in September 1990, led by the first author of this paper. The second is the very large movement called 'Memorial', which has thousands of members and was set up to document the victims of Stalinism. It has been collecting records, photographs, and pictures, and recording interviews, transcribing the interviews, collecting written autobiographies, and so on. But again, all this is so new that, although Memorial is an organization on a scale infinitely larger than any Oral-History Society in the West—an organization with perhaps 5,000 members in Moscow, 5,000 in Leningrad, and 15,000 in Lvov in the western Ukraine—there is still the fear that things will change again. When, therefore, Paul Thompson visited one of the leaders of the Leningrad Memorial in February 1990, and asked where the material was, and was it going to be a public archive, he was told 'Certainly not, how could we possibly trust this material to a public archive?' It was being kept in people's houses, and there was great secrecy about whether there were copies of the material in one person's house, and duplicates somewhere else, and so on: the whole operation was still being conducted in a quasi-underground spirit. Thus, especially if you add the extraordinary practical difficulties of getting equipment or cassettes, and also of publishing, the context of oral history work in the Soviet Union remains at present very different from that in any other major country in the world.

Even so, in the longer perspective contemporary Russian oral history can claim the same roots as the West in the practice of the first great historians of classical Greece, above all Herodotus and Thucydides. The first written chronicles in Russia itself were intended to record medieval oral traditions of the past. And as in the West, the role of oral sources remained important long after the advent of printing. Whether in the law, in the religious traditions transmitted from Byzantium, in popular pagan beliefs, or in everyday custom, they continued to contribute to the development of Russian culture and consciousness. Equally, although the creation of a Russian national history—and with it, a centralized archival system—was a direct reflection of political centralization, some of the best-known earlier national historical writers were as concerned with oral sources as with documents. Thus Golikov, the famous official historian of the reign of Peter the Great, was at the same time a genuine oral historian, who visited the places where people gathered, such as the market-places in many parts of Russia, to listen to the common people. He collected and published a very large number of anecdotes about Peter the Great, which we can read today in thirty volumes.[2] Likewise, Pushkin, perhaps the most famous of all Russian poets, wrote the history of the revolt of the Russian peasantry in Orenburg, close to the Urals, under Pugachev in 1773–5; he collected some 300 interviews with people who could remember Pugachev and details about the revolt. Besides collecting from the common people, Pushkin also kept notebooks of his dinner conversations with fellow aristocrats, which were the source of his well-known *Table Talk*.[3] Some of the reminiscences which he thus collected were printed in the magazine *Russian Archive*, published from the 1860s until 1905, which was at that time very popular among both the Russian intelligentsia and fashionable society. The same magazine also printed eyewitness accounts of the Crimean War, the pogroms against the Jews, and the Russo-Turkish War.[4] There was also an awareness among some major nineteenth-century Russian historians, of both the value of oral traditions in their own country, and the value given to oral sources by the Greek founding fathers of the historical profession. Karamzin, for example, stresses both points in the introduction to his famous *History of the Russian State*.[5] Equally striking was the work of V. Y. Bogoocharsky, the leading historian of the Russian populist movement of the 1860s, who published recollections in the magazines *The Past* and *Past Years*, comparing oral and written sources for reliability, and appealing to surviving witnesses to come forward.[6]

The October Revolution of 1917 brought a brief but remarkable flowering of oral-history activity in the new Soviet Union. Historians tried to collect and write the reminiscences of participants in the 1905 and 1917 revolutions and in the civil war. This was especially supported by the Central Publishing Office for Local History and Local Studies, which pointed out that many ordinary activists did not have the skill to write their reminiscences. It advocated the use of questionnaires, and also the collection of proverbs and songs. Other historians who had left Russia tried to create an oral history of Russian emigration in different countries, for example in Paris. From the late 1920s Soviet local historians began to collect reminiscences on the history of factories and plants, sometimes through speeches at public meetings. Maxim Gorky encouraged the collection of historical material on major construction works such as the building of the Moscow metro. Some of the 'Factory and Plant History' editorial collectives organized factory meetings at which recollections were read out and corrected.[7] To this same phase also belongs the linguistic work of S. I. Bernstein, who built up a pioneering collection—now partly dispersed and lost—of 650 gramophone recordings of talented orators, poets reading their work, and northern storytellers recounting traditional legends.

With the upheavals and chaos of the early 1930s all this oral history work effectively came to a halt. The Communist regime under Stalin paradoxically imposed a historical academicism more oppressive than any in the West. Indeed, the very telling of anecdotes, let alone their recording and publication, became a possible justification for arrest. The only exception of importance right through the Stalin years was the continuing work of Soviet ethnographers and folklorists, who mainly focused on traditional tales. Some of their more recent work, such as studies of continuities and change in burial customs or of changes in family traditions as peasants moved into the towns, seems close in spirit to social history using oral sources.[8] But even the ethnographers had to keep their fieldwork material privately and only publish with great selectivity. They knew of, but could not reveal, the active and often inventive resistance, especially in rural communities, to the new atheistic Communist culture imposed centrally from Moscow. The real Russia remained very different from the uniform illusion created by the centralized media: culturally and socially very isolated—tuneable radios were unavailable to ordinary Russians before the 1950s—and very diverse.

For historians, on the other hand, there was no continuous practice. The experience of the 1920s was revived during the Great Patriotic War, when the Historical Commission organized the collection of some 10,000 testimonies—recorded in shorthand—from both officers and ordinary soldiers on the war front and in hospitals. However, both this material and that collected in the 1920s was left unanalysed. There was thus a long gap in the use of oral sources by Russian historians. In the late 1950s and early 1960s some of the more progressive historians again began to collect popular reminiscences, but many of them were arrested after the fall of Khrushchev. Such work as continued, as in the archive set up by V. D. Doovakin at

Moscow University from 1967, was conducted in semi-secrecy and all publicity was deliberately avoided for the next twenty years. The present change of climate only dates from the beginning of perestroika in the mid-1980s. Even today, Russian oral historians face a strong resistance from academic traditionalists, who dogmatically refuse to give any validity to oral sources, despite the much more positive tradition of earlier Russian historical writing.

The Oral History Centre at the Moscow State Institute has launched a number of different projects, some closer to social history, some more immediately political in their relevance. It started as an informal club in 1985 but in October 1989 became the first official oral history centre in the Soviet Union, jointly sponsored by the Moscow Historical and Archives Institute and the Memorial movement. The members of our centre are post-graduates and students at the institute. The centre undertakes various kinds of activity. First, and perhaps most important, every year, as with the long Russian tradition in disciplines like ethnography, we organize a special oral history fieldwork expedition. For example, in the summer of 1989 we went to the Kuban, a very special part of the Soviet Union, where a particularly interesting national group, the Cossacks, lives. We have everyday work too; all members of the club have their own interest and perhaps twice a week go to interview some old people in connection with this. All these themes relate to the repressions of the last seventy years. We are seeking to interview people who have had any connections with these repressions: maybe their fathers or mothers were repressed during Stalinism; maybe these people themselves took part in the repressions as KGB members, and so on. The centre now has about seventy cassettes containing these interviews.

Let me describe some of our projects briefly, and then look at two in more detail. One of our projects is to interview people who took part in the civil war in Spain. The results were surprising. Officially it was said at the time, and has been generally believed since, that, like those who joined as left-wing volunteers from Britain or France, all the Soviet people who took part in this war were volunteers. However, on interviewing them, we began to understand that many of these people were sent as part of their compulsory military service, and that some were members of the Soviet secret service. They went not to help their Spanish friends, but for their secret work, and rather than killing Fascists, they killed genuine Russian volunteers who were hostile to the Stalin regime. We have also begun collecting histories about the international Red Cross, and as a result uncovered a similar situation: many Soviet doctors, and maybe other kinds of workers, were covert members of the military system; they too carried out their secret tasks. This has been very interesting for us, because Soviet people know nothing about this. Most of these people are unusually difficult to interview, and afraid of their past, but now, perhaps since 1988, some of them have begun to speak openly about their secret work, though, they never say 'Oh yes, it was our mistake.' They say that they worked honestly to serve Soviet interests.

We have also started new projects. In 1990 there was a summer expedition to Samarkand, to interview the Crimean Tartars who were deported there, as part of the forcible removals and dispersal of peasants resisting collectivization in the 1920s. We intend to begin recording in the western Ukraine—the area around the city of Lvov, one of the strongholds of Memorial—which has a particularly complex history, since it was Austrian, and then Polish, before finally being annexed by Soviet military conquest in 1944. There have been long-standing religious conflicts there between Roman Catholics, Uniates (an Orthodox group who recognize the primacy of Rome), and the official Russian Orthodox Church (which although a minority locally, was backed by the Moscow regime)—disputes which have re-erupted since perestroika, bringing an added relevance to the interpretation of recent history.

We also work in Moscow itself. One project focuses on the destruction of many of the city's most famous churches and cemeteries. Russia has always been a notably religious country, but the destroyed buildings included some of exceptional beauty and religious significance. We should like to know who carried out this work, and how it was explained to the people. There is nothing in the official documents which allows us to understand how this desecration was achieved. We are also, in a broader way, trying to understand the consciousness of the older generation. We visit older people and talk to them about the sense they have of the meaning of their lives. We ask them what they think was most important in their life: maybe God, maybe the Communist Party, and so on. We want to compare different generations. We often find that the old people say they believed in God; the middle generation, now aged about fifty, may have believed in Socialism and the Community Party; but the younger generation believe in nothing—or maybe just in George Bush.

Let us look at two of these recording programmes in a little more detail. One of our very first summer expeditions was to Vladimir, in 1984, before perestroika had officially begun. Vladimir is an ancient city three hours east of Moscow by train, briefly the capital of Russia in the Middle Ages, and now visited by tourists for its superb medieval churches and monasteries. Unless you were told, you would be unlikely to realize that up on the hill in front—as you come out of the railway station—concealed behind an ancient fortified monastic wall, was one of the biggest political prisons in Russia, still in use in 1984. We wanted to study the atmosphere around this prison. At this time people in Vladimir were very much frightened to speak with us. But in spite of this, they certainly told us some awful stories about the prison. We found a person who had served in the military at this prison, and it was always possible for us to find people who had been arrested and imprisoned there as political enemies of the socialist State. So we had three kinds of reminiscences: one type was from people who had either sympathized with the repressors, or had actually worked in this prison themselves, another from people who had been captives; and the third, and maybe the most interesting, was from people who had simply lived in the city—some of them said they knew nothing about the prison. But we also found striking differences in the memories collected in different contexts. The first was between those recorded by ourselves and those recorded by local historians. We had come to Vladimir, a provincial city, as Moscovites, and even though we were all young students, people were put off, because to them Moscow symbolized power. They seemed more confident in talking to local historians, who sometimes were also neighbours: they knew them and felt they could trust them. But a still more striking contrast emerged in 1988, when, for the first time, the club formally organized a summer expedition—consisting of a group of seven students sharing two tape recorders. By this time the media—television, newspapers, and magazines—were full of accounts of the Stalinist repressions. When we re-interviewed the very same local inhabitants, their whole historical consciousness seemed to have changed. Not only were they now very willing to talk, but a good many of them—although fortunately others were more consistent and reliable—now also included recollections of real personal experiences as well as 'memories' which they had clearly picked up from the media—perhaps they genuinely believed these had happened to them, enabling them to present a new self-image. Thus some, who previously had apparently not known of the prison's existence, now spoke of themselves as victims of the repression. Others even told us of how they had been arrested and had worked in the Siberian concentration camps.

Understanding this kind of inconsistency is clearly especially vital to Russian oral historians, for we do not have the secure historical framework established by newspapers and records against which most Western historians can evaluate oral sources. It is sometimes said, and is almost true, that 'for us the documents are subjective, and the only things which might be objective are the memories'. The present position is remarkable in another way. The very suddenness of political change, after such a long period in which no contrary views were heard, has thrown the whole framework of historical interpretation up in the air. People find it very difficult to make sense of their own memories without an accepted overall public historical story to relate them to. In 1988 the accepted overview had collapsed: the past was just a set of pieces, and nobody could quite understand the whole. That is partly why you get these instances, as in Vladimir, of people incorporating things they have heard on the television in their own story. They are grasping for stories to make sense of their own lives, and are fitting them in if they can. But the other side is the sheer novelty of talking about such memories. In the Kuban, for example, we recorded the recollections of members of families who had themselves suffered famine yet had never told their grandchildren about those experiences. We found the same thing regarding the memories of the camps. In spring 1989, Memorial organized an exhibition about the camps in the Palace of Culture in Moscow, and for the first time there were maps, documents, and photographs on public display. The thousands of visitors included old people standing there with placards hanging in front of them, saying, 'Does anyone else remember X camp?' They simply wanted to share these personal memories. Others had notices asking about the fate of their parents; others just stood and wept. We are trying to make sense of a really extraordinary situation.

Let us conclude with some more comments on our summer 1989 expedition to the Kuban. Our intention was to record memories of the famine which took place there and also in the Ukraine during the years 1932–3. When we became interested in this theme, we went to libraries and tried to find some books or maybe some research works about this period and this question. But at first we couldn't find any books by Soviet historians about that question; we could find only the books of Western historians. Then we decided to look through the newspapers which we had in the Soviet Union at that time, and we were greatly surprised when we discovered that even in the newspapers there was no information about the famine. There was a lot of information about Stalin and the prosperity of the Soviet State, and happy news about collectivization, industrialization, and the victory of Communism, as well as news about capitalism and its failures. But there was no information about the famine. The official records are still not accessible. Memorial had asked to use them, but were refused. They therefore encouraged our oral history expedition. We wanted to use the memories of local people to try and understand whether that famine was deliberately organized by the Soviet State, or whether it was natural, maybe because of bad weather? When we began planning our expedition to the Kuban, we learnt the history of the population of this territory. The population consists of Cossacks, people who for many centuries lived on the boundaries between Russia and states like Turkey and the Tartar kingdom. For two hundred years they had been soldiers. That was why they had always been able to withstand a strong influence from Moscow on their way of life, and why they had such strong feelings of freedom. Maybe freedom was the most important feature in their national character.

The years 1930–3 were the crucial years of Soviet industrialization. The Moscow government at that time needed a lot of capital to buy machinery from Western countries, and consequently had to find as much hard currency as it could. Alternatively it could pay directly through exporting grain. The State therefore needed as much grain as it could get. But the peasantry could not survive if it gave up all the grain it produced. This becomes clear when you look at the different interpretations of the word 'sabotage'. This is a word much used in official Soviet history texts. We wanted to know what it meant on the ground. When we began asking people, they told us, 'Oh, sabotage? Sabotage, it was the time when the State didn't give us any food, any grain, and we had to suffer famine.' But at the same time the official press and the official historical record describe sabotage as the time when the peasantry hid all their grain, and refused to give it to the State, which was why there was a danger of hunger. We wanted to know what had been the reality. So, it was very interesting to find, at another level, that the local official bureaucracy had a double position. On the one hand, they understood and knew the real situation in the villages of the Kuban, and they realized that the peasantry had nothing more to give to the State. On the other hand, they gave distorted information to Moscow, reporting that the peasantry had a lot of bread, a lot of food, and a lot of everything. As a result of these positive reports, Moscow decided on repressive measures to seize the grain. Thus, according to all the textbooks of Soviet history, we knew that the 'kulaks'—the more prosperous farmers—hid tons and tons of grain. Then we asked people where they had hidden it at that time? And they told us that when the hunger began—at that time when the famine began, which was after the collectivization of the land—they had no grain at all. Everything had been taken from them, from their houses. They had only a little food, enough for themselves and for their children. For example, one woman told us that when some soldiers of the internal forces came to her house, they wore red armbands and carried special sharp sticks, with which they poked under the ground and in the beds of people. But they could find nothing in their houses, so that when they saw their last pot of potatoes on the table, they took even the potatoes. It was a situation when people were hiding not tons and tons of grain but just a little something: maybe some pieces of food for their children, or a pot with porridge, vegetables, or something like that. So that was the reality in the villages.

One old man told us of an experience which we did not understand at first, namely that his village was put 'on the black board'; not on the black list, but on the black board. We began to ask people what this meant, because it was strange to us. They told us that in some Cossack villages state officials thought that people might have some food, grain, or bread, or were afraid of resistance from the population so they put these villages 'on the black board'. This meant that instead of the red banner which was flown near the office of the agricultural Soviet in every village, they put out a rag that had been dipped in black tar. It looked like a black board, because under the weight of the tar it could not move. It meant that this village was surrounded, under a military commander, and blockaded by troops, and nobody was allowed to leave it. This was why in such villages especially, a lot of people died. The official forces took all the grain and other food from the people's houses and did not supply them with any food at all. Nor did they give them any opportunity to find food in villages or in nearby cities.

Different perceptions of the past also came out very strikingly in the implications of the team 'red partisans'. We were surprised when we were told that together with the military, red partisans took part in this blockade of the villages. So we began to ask, 'why "red partisans" ': it was not a war, it was not the Second World War, why 'partisans'? And they told us that was how those who helped the military to blockade the village were called, because of their red armbands. Certainly, this had a profound impact on the historical consciousness of people. That is why, when the territory of the Kuban was conquered by the Nazi troops during the Second World War, at first the people didn't support partisans, who struggled against German occupation. For the Cossacks the meaning of the word 'partisans' was connected only with famine. They thought that partisans were people who helped Fascists. Everything was so mixed up in their mind and in their consciousness. It seems that the confusion of the symbolic meaning of red goes back to even before this. In retrospect it is very difficult for us to understand why people became red partisans during the famine. For example, one man served for about two weeks as a red partisan. When he returned to his house in the village, he found that all his own family had died because of famine. Yet, he stood guard on the road, and he refused to let other families, other people, leave those villages, despite his own family having died of hunger. We also discovered that there were some people who helped the soldiers look for grain in the houses of others. We asked why people, including women too, did that; why they took part in this awful thing? When we enquired whether the red partisans were the same as the troops, why villagers should join the partisans, and asked 'what was the difference?', one woman told us, 'Oh, but they had the red bands on their arms.' When we asked, 'Oh, what does it mean?', she answered, 'Oh, it was so beautiful!' Such an answer is certainly very difficult to understand, but it seems to go back to traditions from the Civil War of 1918–21, because in the Kuban there was a very difficult situation then, with a lot of battles waged between the Red Army and the White Army. Villagers who had sheltered one army were liable to be shot when the other arrived. So, when the Red Army was in the village, women tried not to wear coloured dresses, because if you had a coloured dress, the soldiers of the Red Army assumed that you were a 'kulak' and came from a rich family: hence, they tried to put on some dark dress, or black dress. But when the White Army conquered these villages, women immediately tried to change that dress, and to put on jewels and all their clothes with bright colours. Hence, after the eventual victory of the Bolsheviks, many women never had the opportunity of putting on their bright dresses again: they only dressed in dark colours. Of course, in reality, they didn't have many dresses at all. That is why when they had an opportunity to receive some garment in colour, even red armbands, it was a special pleasure for them. It made them happy. It seems that this was one reason for their putting on those red bands, going to their neighbours, looking for the grain, and taking food from the children.

Consciousness, it seems, was as confused in the past itself as it now seems in memory. We did not find a common collective memory in the Kuban. Individuals structured their answers in very different ways, so that interpretation was always a complex and difficult task. We found that even in the same village, there might have been starvation in a Cossack quarter, while in another part there were ex-soldiers from Leningrad who always had some food: yet, neither understood the experience of the other. What we heard was essentially a direct account of experience which was rarely understood in context. Some of the most isolated villages had never known there was a widespread famine at all. We would simply ask the older villagers general questions to draw out their life stories. It was often the first time they had recounted their experiences, which were frequently horrific and included even cannibalism, to anyone. They had remained terrified of further sufferings, including imprisonment and deportation. They did not even speak of the past in their own homes, and had not told their own grandchildren. The stories they told us were still uncomprehended, uninterpreted, unmitigated.

It is still not easy to carry out fieldwork of this kind, even in the Soviet Union today. In the Kuban we still found that many people would refuse to be interviewed and say, 'No, I'm not going to speak. In case things change, I'm not going to say anything.'

When we began our work on the Kuban, people had not had any opportunity—for they were frightened to death, to real death—to speak about their experiences for more than fifty years. It was very difficult for them to recall their past, and to be reminded about it. A lot of people were shocked when we asked them to tell us about the famine. In the official histories and propaganda, it had been maintained that there had been no famines at all. So, a lot of people, especially at first when we were still strangers to them, told us, 'Oh no, I don't know, I know nothing.' Many others told us, 'Oh, it was such an awful life, it was awful. So I don't have the strength to remember about that. I can't speak about that.' The majority of people who did speak, cried during the interviews. It was impossible even for us to listen to maybe four or five interviews a day, while all the time seeing people crying. It was awful; psychologically, it was very hard.

The Kuban is an extreme case, but it is by no means unique. Even in

Moscow itself, when we want to interview someone, people very often ask us 'Who are you? Why are you interested in all these questions? Have you any identity document?' We say, 'Oh, there is no problem. Everything is OK, I am a student from the institute, I'm not from the KGB.' And they say, 'Oh, but maybe you have two documents! In this pocket and this pocket!' One from the university; the other from the KGB . . . Yes, from the start to the finish, oral history work in the Soviet Union is no simple undertaking.

Notes

1. For example, Leopold H. Haimson (ed.), *The Making of Three Russian Revolutionaries: Voices from the Menshevik Past*, New York, 1987; or the questionnaire-based survey of the Soviet Interview Project analysed in James R. Millar (ed.), *Politics, Work and Daily Life in the USSR*, New York, 1987, a useful but unfortunately lifeless collection of essays.
2. Ivan Ivanovich Golikov, *Delniya Petra Velykoso* (The Deeds of Peter the Great), 30 volumes, Moscow, 1788–97.
3. Alexander Sergeevich Pushkin, *Pugachevchina* (About Pugachov); *Zapisniye Knishkeye Knyagineye Zagryashskoy* (Notes of the Lady Zagryashnaya); and *Table Talk*, Berlin, 1923.
4. *Russky Archiv*, under the editorship of P. I. Bartenev; see also *Russkaya Starina* (Russian Antiquity), edited from Petersburg by M. I. Semevsky. Another earlier precedent is V. Siroechkovsky, *Moskovskiye Slukhy* (Moscow Rumours), Moscow, 1825–7, based on the strange tales of night-watchmen.
5. Nikolai Mikhailovich Karamzin, *Istoria Gosydarstva Rossiskovo*, 12 volumes, St Petersburg, 1842.
6. *Billoye* (The Past); *Munuvshye Gody* (Past Years).
7. *Istoria Fabrik y Zavodov* (Factory History), 2 volumes, Moscow, 1935; *Kak me stroily Metro* (How we Built the Metro), Moscow, 1933–4; *Kak me Spacaly Cheluskintsev* (How we Saved Cheluskintsev—the North-Pole Explorer), Moscow, 1933.
8. Margarita Nikolaevna Shmeleva and S. P. Boodina, *Gorod y Narodnye Traditsye Russky* (The Town and Russian Popular Traditions), Moscow, 1989.

6

The Gulag in Memory

IRINA SHERBAKOVA

The uniqueness of our Soviet situation lies in the decades through which historical truth within our country lived on only through underground memory. Publications which appeared in the West, broadcasts from foreign radio stations, and manuscripts secretly circulated in samizdat form only began to seep through comparatively widely as recently as the 1960s and 1970s. Before that there was only silence, or at the most, rare whispers between intimates, because to tell anyone about the prisons and concentration camps was deadly dangerous.

I began collecting accounts of the Stalinist camps myself in the mid-1970s. At first I only recorded women whom I knew well as close family friends: often I had listened to their stories since my childhood. Gradually the circle of people I was recording widened, as I was sent from one person to another. I started to use a tape recorder in 1978. Altogether, I have now recorded 250 ex-prisoners. Sometimes I have listened to someone for months, but with others I have only talked for a few hours. The interviewees fall into several groups, but the largest number are 'victims of 1937', imprisoned at the height of the terror, followed by those convicted in the renewed repressions of the late 1940s and early 1950s. A much smaller group of my interviewees started their prison lives in the 1920s. The first of them all was imprisoned in 1919, and the last in 1953. Two-thirds of them are women, which may well be a typical proportion of survivors generally. In terms of nationality, the majority are Russian, one-third are Jewish, and the rest are Polish, German, Latvian, Armenian, and so on. As to social background, the majority come from the urban intelligentsia—doctors, teachers, scholars, students, or journalists—while others were party workers or from the armed forces.

For the Soviet regime, memory itself was intrinsically a serious threat. The entire history of the past, and above all of the revolution

Translator: Paul Thompson.

and the civil war, was rewritten and mythologized. Memory was made to function in a truly Orwellian style: what had been peace was now declared war, and the Soviet version of memory became oblivion.

The time span of many decades, compared with just twelve years of Nazi power in Germany, has shaped the character of memory and recollection. Those who fell before the waves of repression in the 1920s—beginning with members of the so-called 'exploiter classes', former aristocrats and White officers, followed by Social Democrats and social revolutionaries, and later by Trotskyites—were, as a rule, unable to emerge from the Gulag system before the 1950s. My interviews include many of this kind. For example, Daria Samaelova was first arrested while still at school, at the age of seventeen, on the grounds that her older brother had also been arrested as a Trotskyite opposition sympathizer. This was in 1927, in the city of Baku in the Caucasus; she was released after two months. But a year later she was arrested again and deported for three years. In 1933 she tried to re-enter normal life and, hiding her past, enrolled as a student in Moscow University. But after a year she was rearrested and exiled to Tartaria. Then in 1936, she was once again arrested, this time receiving a prison sentence of ten years for Trotskyist counter-revolutionary activity. In 1946, after her release, she was permanently exiled to Kolyma (in the far north-east of Siberia). Samaelova was only rehabilitated after 1956. And far from being a unique case, hers was a characteristic path which hundreds of thousands of others followed. It is unusual only in that she survived such a past to be still alive today, so that I was able to record her history.

Obviously it is impossible to repress the memory of such a fate: that would mean forgetting one's entire life. The effect is rather different: more of a confusion—especially in the memory of prisons and camps—of the superficial distinctions between imprisonment and freedom. Today they all repeat the observation, which is now a commonplace, that the Soviet Union in the mid-1930s was itself an immense concentration camp: even if millions of its people, particularly of the generation born after the revolution, considered their 'cell' the brightest and most beautiful in the world. Thus, those who had served their term and were released into freedom still felt—either consciously or unconsciously—that the Soviet Union itself was a concentration camp, and therefore chose not to return to the 'big world', but instead settled down in Siberia or the far east of Russia, living near the camps, continuing to work on the camp farms on different tasks, as exiles. The most far-sighted of them did understand the system. But where could they go, where else could they settle? Ex-prisoners were forbidden to live in the big cities, and in smaller towns a stranger would always be under suspicion. Their fears were sound, for a new wave of repression followed in the 1940s, and those so-called 'recidivists' who had settled down in new places were rearrested and sent off into exile: an exile which is often remembered as scarcely less terrible than the concentration camps, with no work, nothing to eat, and nothing to provide warmth. Indeed, those who were sentenced to the camps in the post-war period—with the exception, of course, of those sent to hard-labour camps—often told me, 'You were better fed in the concentration camps than in freedom.' The camps formed a network spread over Siberia, the far East, Kazakhstan, and so almost the whole country, which was essential to the immense Soviet industrial and agricultural projects, supplying a labour force which therefore had to be fed. Life was certainly tough in these camps, but their objective, in contrast to many of the camps in the repressions of 1937–9, was never annihilation through mass executions or death through starvation. The prisoners in the post-war camps adapted themselves to these conditions, and prepared to spend their whole lives in the camps, to which they had already been typically sentenced for ten to twenty-five years. When release came, after the death of Stalin, it was neither anticipated nor easy, and this is partly why there often is a lack of clarity and coherent perspective, and in some respects confusion, in their reminiscences. For them freedom did not come in a single, swift, joyful act. Release demanded trouble, letters, and petitions. Re-entry into ordinary life was slow. They had long struggles to win rehabilitation, a flat, or a pension. In the Soviet situation it was very difficult to perceive the end of repression in an individual's fate, not merely in memory, but simply in real life. For many of them the repressions have scarcely finished even in the most recent years, when they have at last been able to talk openly about their own pasts.

What and how did they remember and recollect, thirty or fifty years ago—and today?

It may sound paradoxical, but we in Russia today are certainly persuaded of how short the span of human memory can be. In the last three or four years we have so thoroughly examined—as well as watched on the screen and heard on the radio—so many reminiscences of the repressions, the prisons, and the concentration camps of the pre-Stalinist, Stalinist, and post-Stalinist periods, that very quickly it suddenly seems as if this past was always known and remembered. Yet a mere ten or fifteen years ago, when the events of the 1920s–50s were much closer and their witnesses younger, it seemed that everything of the past had disappeared into oblivion, and that living witnesses were lone individuals: for they were still in deep concealment.

The first wave of recollections of the camps and prisons poured out during the Khrushchev thaw, when the first ex-prisoners began to talk about their experiences. Those who came back can be roughly divided into two groups. There were those who wanted to forget what had happened to them. When I first began my recording in the 1970s, I encountered, among the families of ex-convicts who had married and had children after their release, some members who did not know about their wife's or mother's past. But there was another group who right through their imprisonment, sometimes subconsciously and sometimes explicitly, had wanted to remember. A determination had developed in them that what had happened to them and those with them should be fixed in the memory and perhaps later recounted: and this had helped them to survive. Many of those who wrote their reminiscences in 'fresh footsteps' in the late 1950s and early 1960s spoke about this. They often took the attitude that if the chance to recollect was missed it would be irretrievably lost to us. Perhaps we can recall the recording of the stories—the fresh stories—in the late 1950s, of those who were waiting in queues at the military prosecution offices to obtain their certificates of rehabilitation. At that time there were some among those returning who undertook Moscow 'lecture tours', orally recounting camp experiences, because interest about the Gulag was so high among the intelligentsia on the other side of the barbed wire. But, in literary form, the facts about the prisons and camps were as a rule tragi-comic. In particular, this was the time when certain expressions and terms from camp jargon began to pop up constantly in conversational language. Nevertheless, the real picture of the repressions and of prison and camp life remained entirely hidden. For beyond imprisonment glimmered a ghastlier fantasy. The reality of the camps had been terrible and fantastic enough, but the whole horror was submerged in the incredibly depressing and petty everyday struggles for survival. This was why the appearance of Solzhenitsyn's novel *One Day in the Life of Ivan Denisovich* seemed so important. I well remember how, reading it as a schoolgirl, I was above all puzzled: I could not understand what had produced such a powerful impression on my parents. I had expected something horrific, freezing the blood in my veins, but all that Solzhenitsyn offered was hunger, cold, filth, and unmanageable labour, a mere biological existence, in which every rag and every tiny crumb of bread took on an existential significance. But ex-prisoners at last saw a rendering of part of their own lives in the novel, and a stream of recollections poured out with the journal reviews. However, the camp theme was very soon officially prohibited again, while official biographies began to use formulas to symbolize that a person had suffered under the repressions: 'In the 1930s and 1940s he worked in the far North', or 'his life was torn apart in 1937'.

But by now it was already impossible to force all the witnesses to subside again. The books of Solzhenitsyn and the recollections of Evgeny Ginsburg and Varlam Shalamoff were reprinted and broadly circulated in samizdaty. Above all, some ex-convicts were asserting the need to talk about their past and of being together with other survivors. Quietly, those who had been in Kolyma, Vorkuta, or Inta were meeting. Nor was this surprising. Among those repressed under the infamous political Decree 58, especially in the 1930s, were many active city intellectuals. It was well known that in almost every concentration camp a university could have been opened. Links made in the camps were often maintained in liberty. People met together and continued to reminisce or even read aloud their testimonies, which had been introduced as court evidence.

On the other hand, to speak to outsiders brought a fear, a fully justifiable fear, that everything might be repeated. With my informants in the 1970s, I constantly used to find that they became instantly silent on seeing my tape recorder. An additional hindrance was their memory of the false confessions, obtained through physical and psychological tortures by the NKVD—the People's Commissariat of Internal Affairs, the internal security police—which were preserved somewhere in the archives. Only now, after having at last got the opportunity of getting to know some official investigation papers, including the minutes of interrogations, and reading the incredible evidence from their nearest and closest—husband, lover, or friend—have I finally been able to confront what forced victims either to keep their silence or to expunge whole sections of their consciousness. Recently, an archivist told me the story of a woman who applied to him to be shown the papers on her father, who had perished in the late 1930s. When he looked at this evidence he discovered that the unique accusatory material was that of this very daughter, who was then aged sixteen. They had taken her to the interrogation and, already frightened by this, she had used some careless phrase of her father, in an argument with him—just letting off steam—the previous evening. But afterwards, of course, she did everything to put this out of her consciousness—and indeed she 'forgot'.

But the perpetually recurring question was how those who were strong and brave people in ordinary life came to denounce themselves and others. The limit of endurance was very plainly explained to me by Ivan Fitterman, an important constructor at the largest automobile factory in the country, who was arrested in the 1950s and sentenced to twenty-five years as a participant in Zionist plotting. He was in the notorious Sukhanovskaya prison, within 40 kilometres of Moscow, which was intended to destroy all who persisted with resistance: 'If they needed to get evidence from you, they got it. Or they killed you. If you did not sign what you were given, you knew that the moment they waved their hand at you, you might be no longer necessary for their evidence.'

'The investigator responsible for our rehabilitation', recalled Irina Kin, the widow of a famous Soviet journalist shot in 1938, herself imprisoned for ten years in the camps, 'showed me documents about my husband. I knew his handwriting; I read how he denounced himself—and I began to be hysterical. But he said to me, "I thought you were a strong woman." '

For many years people found it hard to understand why the punitive administration so persistently sought these confessions. Probably there is one explanation: that in a totalitarian society in the period of total terror, the general guilt needed to be shared by all—by both victims and executioners.

In general, the problem of who were victims and who were executioners in the Soviet context is extraordinarily complex. It is not a myth that those who were shot sometimes died with the name of Stalin on their lips, maintaining to the last minute their belief that their beloved leader knew nothing of the repression. Those arrested in the early and mid-1930s included a high percentage of believing communists. Many even returned from the concentration camps as convinced communists and rejoined the party, by no means merely for the sake of their pension and privileges, or for an easier way of fitting into normal life—although that was one reason—but because they continued to believe in the excellence of communist ideas, explaining their own misfortunes in terms of mistakes and distortions, and also continued to believe in the personal virtues of Stalin. Lenin remained a sacred image, and all that was necessary, as they would then say, was to return 'to Leninist lines'. The ironic consequence of the Gulag system was that both victim and executioner could share a common ideological platform. Neither the investigators, nor the guards or the administrator of the camps could be ideological enemies for arrested communists. Hence, the first thought of those arrested often was that the arrest must be a mistake, a misunderstanding; that the investigator was not an enemy, like the Gestapo, but a Soviet man who simply needed to be convinced that you were not guilty. It happened in some cases that the NKVD investigators turned out to be acquaintances whom they had met in general friendship, and who by no means turned on full moral or psychological pressure. 'There look, but if we had behaved ourselves well, we could be drinking tea together with you at Mr H.'s', recalled Olga Penzo, a ballerina arrested in 1937.

Hence some of those things seem wholly incomprehensible to someone of a different generation: like those pages in the album of an ex-convict, where I noticed among the photographs from the late 1950s a snap of herself with other women at one of the Crimean health spas. One of these women, it was explained, had been the governor of the camp in which my informant had spent some years. Already in the Khrushchev period they had met again in the spa and were photographed together as long-remembered acquaintances. 'She wasn't the worst from there', Katya Gavralova, who had spent eight years in the Kolyma camp, explained to me—'there were some much worse than her'.

Now, there are former camp administrators who write letters to ex-prisoners asking them if they will confirm, for their personal records, that they were decent and discontented with the party organization. And the ex-prisoners reply: 'Yes, he wasn't so bad, at least he didn't steal from people, or starve them as much as the governor of the next camp.' It is important to note that among survivors, a large percentage survived through being in relatively privileged situations, working in the administrative structure close to camp authority.

None of this is surprising: it was similar in the Nazi concentration camps. The difference was that in the Soviet camps the victims were often identifying themselves with the very regime which was inflicting their repression. Thus I have often heard such views expressed as: 'When I was in the prison or camps I was ashamed in front of young women students, or foreigners, about what was happening in my Soviet country . . .' After the war it was still shameful in front of Latvian women, Estonians, or west Ukrainians. And the source of this shame was in their genuine loyalty to the regime, even in the camp or prison. This is the striking meaning behind the apparently absurd remark in the story of a German woman who came as a political immigrant to the Soviet Union in the 1930s and was arrested there: 'How glad I was that I could participate in the building of socialism at Kolyma, how much more incongruous it would have been for me to have served my sentence in some Nazi prison . . .'

Sometimes the links between the two sides were superimposed quite fantastically. Tamara Galkina was the wife of an important military man, but she fell in love with a young subordinate and went off with him. He had an active career in the NKVD. In 1937 they came in the night but, to their surprise, they did not arrest him, but took her—to her first husband, who was being held in connection with the Tukachevsky case. Her first husband Mikhael was shot, and she herself was sentenced to ten years in the camps. Meanwhile her second husband went into the Gulag as a high-up administrator and at the moment of her release he arrived in Magadan, the capital of the Kolyma region, as a general, one of the chief administrators of the immense network of camps in the far east. He summoned Tamara to him. She knew that after her arrest he had given evidence against her, refuted her, and then divorced her. He now proposed that she should stay with him. She refused. But years later, after her rehabilitation, when he was already retired, they were reconciled, and right up to his death they regularly called on each other, and drank tea together like good old friends . . .

However, this NKVD general was an exceptional success among those who were arrested and those who were shot. In the 1930s tens of thousands of the NKVD workers themselves perished. They led the repressions from the start; they prepared the process of fabrication of documentation; and then they imprisoned themselves. There were many occasions in later years when the sometime tormentor stumbled on his victim in a camp where both were imprisoned.

In the Soviet situation it was possible for a man to change his life role several times: to shoot White officers in the civil war, to participate in the crushing of the Kronstadt rising, then to struggle actively against the socialist opposition, destroying social revolutionaries and Trotskyists, and later to lead collectivization and the breaking-up of kulak farms with an iron fist, and finally in 1937 or 1938 to become a victim of repression himself. As late as the 1970s, I still often heard ecstatic and uncritical recollections about participation in the civil war, or in collectivization: everything was good and proper, until they were struck by the sudden blow of their own arrest.

There are some questions which I always put to my informants: at the moment of their arrest, how far did they understand what was going on in their country; when were their eyes opened, if they were opened—on arrest, in prison, in the camps, or even later? Very few of them correctly appreciated what was happening because, as a rule, the earliest people who found themselves under threat, either because of their background or because of their political beliefs, were already overrun by the gathering speed of the machine of terror in the 1920s. The least prepared and most blind were the communist victims of 1937. 'It seemed an inexplicable nightmare', I heard again and again, or 'The only thing which came into my head was "It's a Fascist conspiracy." ' 'I scarcely understood anything', recalled Dina Yankovskaya; 'as a scientist, a biologist, when at the beginning of 1939 by a happy chance I found myself released from prison, I threw myself into writing letters to the government: to confirm that all those people in prison with me were innocents, and that the NKVD had been infiltrated by Fascists.' The only surprising thing about her story is that rather than immediately rearresting her, they sent her son to the battle-front. She herself survived the Leningrad blockade, before the authorities, after nearly ten years, recollected her letter and arrested her.

'I understood nothing at that stage,' a Comintern worker, Georgei Rubinstein, told me, 'so that after my arrest my first thought in the Lubyanka prison was "They're testing me through a routine task." '

Not surprisingly, nor deliberately do interviewees evoke how far what was happening in the prisons recalled the theatre of the absurd. The wife of a well-known Soviet aviator, who, like him, was arrested, was forced during her interrogation in the Butyrskaya prison to write her husband a letter just as if she was at liberty: everything was in order, she was well and active, the children were studying, and so on. Other arrested women managed—from their cells—to change their clothing, have their hair done at a hairdresser's, and arrange to see their husbands, who also needed to think that they were free. Understandably, they remember the details of such absurd episodes throughout their lives.

In recollections of the Gulag there is always a great significance in the detail, in episodes, in the fine particulars of life: both because of its continuous meanness and equally because it was on this that life itself depended. As a rule, it is the first arrest or the first cell, which is remembered best of all, probably because it came as such a powerful shock to one who, up to that very moment, had been an innocent person. Often, they still hold the whole view of what they saw in their parting glance as they were marched off, their last look at their old life: a handful of sweets scattered on the table, the teapot on the cooking range, the linen in the washtub, the unfinished report on their last work trip which had seemed so important, or the kind neighbours who had managed to return a borrowed coffee grinder just five minutes before the arrest. Each spare little shred of clothing, the stockings given to someone, the mittens stolen from you, they all remained fixed in the memory for decades.

With the millions of Soviet people who had gone through the camps scattered across the whole country, I never cease to be surprised by the part played by unexpected encounters, meetings, intersections, or coincidences: how a man saw someone, how a woman was in prison with someone else. For example, I remembered an episode from Evgeny Ginsburg's beautiful memoir, *Life's Hard Path*, set in the Butyrskaya prison: 'People peered eagerly into their faces—who were they? What about those four, for instance? In such absurd evening dresses with exaggerated *décolletés*, and high-heeled slippers . . . Everything crumpled, pulled at . . . That's the guests of Rudzutak. They were all arrested at his house as his guests.' Jan Rudzutak had been a major political and party activist and a Politburo member at the moment of his arrest. What had happened to these women who flashed past in the pages of Ginsburg's book to be devoured by the vast camp system? And then, fifty years later, one of these same women was sitting in front of me; she had been arrested, with her husband, in May 1937 in Rudzutak's dacha. Her appearance in the prison had certainly been absurd. 'They didn't allow us to take in anything—our summer silk dresses and underwear were torn, stuck over our shoulders, our stockings were ripped. At the last minute before our prison transfer I received money from someone and we bought whatever there was in the prison chest.' So an orange top-shirt replaced a scarf on the head, a green football outfit, trousers, and on her feet the high-heeled slippers disappeared and towels were wrapped round. 'This was how I started the journey in December. It was a pure miracle I survived.'

Over many decades, life in the Gulag gave birth to endless rumours, legends, and myths, most of all of course about famous people—long previously believed executed by shooting in Moscow—who were said to have been seen by someone in some far distant camp somewhere. There were constantly recurring themes and details in such stories. For example, at least four women described to me exactly the same scene: how, many years later, when they were able to see themselves in a mirror again for the first time, the image they saw was the face of their own mother. Even as early as the 1970s, I myself recognized incidents recounted to me orally which were also scenes described in Solzhenitsyn's *Gulag Archipelago* or other printed recollections. But now story-telling about the camps has become so general that recording oral memory has become much more difficult. The vast amount of information pouring out of people often seems to happen through an immolation of their own memories to the point where it begins to seem as if everything they know happened to them personally. As a result, while earlier it had been especially important for me to learn details of their lives in the camps and prisons, now it became much more interesting to catch the 'alienation effect' of being at the same time both witnesses and questioners, of how they were themselves fifty or forty years ago. Especially recently, the moment has come for many—although certainly not all—to make a clear distinction between their present and past selves. The painful wound of semi-rehabilitation and the continuing necessity of silence over long years has now left them. On the contrary, ex-prisoners frequently perceive themselves as bearers of important information. The pain of the past recedes and they—especially the women—can talk about the torture of the prisons almost as if it had not been their own. An English-language teacher, Dina Slavutskaya, after describing to me how she had been forced to stand continuously for three days and nights during her interrogation—a form of torture in which she was forbidden to sit or even lean, and through which she had become lame for her entire life—at the end of her story asked: 'but have I told you something dreadful enough?' 'When I saw the prisoners in Kolyma sifting gold,' another woman told me, 'I vowed to myself that if I survived and eventually came out, I would never wear a single gold piece—but now look at me: look at my ear-rings, my chain, my ring . . .'

So far, I have not referred to the continuing contrasts between the stories told by men and those told by women. As a rule, women describe in much fuller detail the everyday routine of the prisons and concentration camps, the clothing, and the appearance and character of people; they speak more calmly and with more detachment about the tortures and agonies they had to bear; and they are more frank about emotional and sexual life in the camps. Indeed, love in the camps is still one of the most painful and difficult themes to speak about. Women see themselves as generally better able to adapt to imprisonment and they much more often stress the crucial importance of friendships and human relationships in this. They also give more emphasis to family relationships: above all to the tragedy of being parted from their children. Men, by contrast, are more reserved. They more often highlight a moment of personal humiliation. They are more often analytical. Men also usually give more detailed and elaborate descriptions of work, occasionally illustrated by drawings; and they sometimes have a better memory for the names of people they met, and for dates. Hence sometimes, although not always, it can be useful when a man and a woman exchange memories together: a husband and wife, if both are ex-convicts, can add to and correct each other's accounts.

Some of the memories which once were the most painful of all have often now softened. In the 1950s and 1960s, for example, the problem of informers remained very acute. There was of course an immense multitude of people who had helped with the arrests and who had become interrogators in the prisons and camps. Those who returned in the 1950s could very easily meet up with the people who had put them in prison—at work, as guests, or on the street. It was extremely difficult to prove who had informed on them and usually any attempt would only end in public scandal. A strange process is now taking place, one which can be illustrated by an incident which I witnessed myself. At one of the Moscow meetings of the Memorial society, which brought together ex-convicts with those who were studying the history of the repressions, one woman, who had been arrested in about 1939, said to me in a completely calm voice: 'but over there is the man who informed on me.' And she greeted him quite normally. Catching my perplexed expression, she explained: 'Of course we were then just eighteen, his parents were old Bolsheviks who were repressed, and then they tried to recruit me too. And of course he was repressed later on himself.' I felt that what she said was the outcome not of a lack of concern for or forgetting of her past, but of a realization, which now at last had come to people, of the shameful things which the system itself had done to them. In such a situation people need to behave in their own different ways but it is very important not to presume that somebody still carries the same moral self-estimation they once bore in the past. It was much more difficult in the past than today. Now survivors, or at least those who have not lost the ability for self-reflection, are often changing their own self-evaluation. 'How could I have agitated for those unlucky women to go to the collective farms?' one of my women informants castigates herself: 'I never asked myself that, when the KGB came for me—I thought they were arresting me again, but instead they informed me of the death of my brother, for whom I had looked for years in the camps—my first thought was not about him, but about myself. I thought, "Thank God, they didn't put me in . . ." '

The question now often arises, especially from younger people, of the possibility of opposition in concentration camp conditions. It is well known that there were of course major forms of direct resistance in the post-war period, including armed escapes, and eventually after Stalin's death uprisings and strikes in the camps themselves. But it seems to me more important, both from my discussions with the 'last' true witnesses now, and also from talking with them in the 1970s when they were younger, that ultimately the only true form of resistance was a determination to conserve one's humanity. It was sometimes undoubtedly very difficult not to become a camp of wolves, faced with the pitiless law of camp life: 'You die today, I die tomorrow.' And, together with the survivors themselves, we do have now, at last, a final chance of evaluating their past journey, with neither partiality nor anger.

7

The Abduction of Imre Nagy and his Group

The 'Rashomon' Effect

András Kovács

On the morning of 4 November 1956 the inhabitants of Budapest, the capital of Hungary, were woken by the rumbling of Soviet tanks. The armoured troops besieging the city set off before dawn to trample down the first anti-totalitarian revolution of the post-war era. A few hours later a group of men, women, and children entered the building of the Yugoslav embassy. Five members of the leading body of the newly organized Communist Party—Imre Nagy, the revolution's Prime Minister, György Lukács, the philosopher of world-wide renown, together with their closest colleagues and their families—accepted the offer of the Yugoslav government to grant, if necessary, political asylum to Imre Nagy, the members of his government, and anybody else considered worthy by the leaders of the revolution. The twenty-six adults and thirteen children remained at the embassy until 22 November 1956. After a number of negotiations, and because Imre Nagy refused to leave the country of his free will, the Kádár government installed by the Russians promised the group free withdrawal. But the bus they sent to the embassy to 'take them home' drove the group to a military barracks on the outskirts of the city. That was the place where the long tribulation of Imre Nagy and his colleagues began. Some of them died on the gallows; others were imprisoned or exiled for many long years.

Historians have for long shown a great interest in the story of Imre Nagy's abduction. Who made the decision, and how? These questions are important not only from the point of view of subsequent political events in Hungary but also because the answers could cast new light on some of the less explored areas of international politics, and might help to explain why the relationship between the Soviet Union and Yugoslavia changed again in that period.

Only very few written documents have been published about this event. Historians for the most part must depend on recollections and oral accounts. The story of Imre Nagy and the other prominent figures of 1956, a dramatic chapter of Hungarian contemporary history, has become one of the most important focuses of research for Hungarian oral history in the past few years.

The investigation of the factual course of certain especially significant historical events is only one of the possible aims of oral history, though perhaps the most difficult of all. In the following account I would like to explore the difficulties which may arise in the process of reconstructing such highly significant historical events through oral history, using as an example the oral accounts of the abduction of Imre Nagy and his group.[1] I have selected one sequence from the available material covering a well-defined period of about two hours, beginning with the group leaving the embassy and ending with their arrival in the barracks. This short period seems especially suitable for the purpose of our analysis because the circumstances obliged everyone to be an 'observer' of everything that happened, and judging by the narratives, it made a very strong impact on the memory of those involved, with the result that they are able to recall even some very fine details in their narratives.

At the start of the analysis, I tried to create a story with a 'common denominator', constructing it from the elements which recurred in each narrative or where none was contradictory. The result was surprisingly sparse in detail.

They left the embassy sometime late in the afternoon. Even before getting on the bus there was uncertainty among the members of the group. More and more suspicious circumstances were revealed as they got on. Not only the bus itself, standing in front of the building, but also the driver was Russian. A short discussion took place between Imre Nagy and Soldatić, the Yugoslav ambassador. The final decision to leave the embassy building was made after this. The Yugoslavs designated a few escorts. The bus drove off. In a few minutes it became obvious that the passengers were not being taken home since the bus did not turn into the street where Mrs D. lived, only a few hundred metres from the embassy. 'How can they do such things to grown-up people', this elderly and completely apolitical woman, the mother of one of Nagy's closest colleagues, said indignantly. Confusion became complete in the bus. Then they stopped. A few escorts got off.

Some new unknown ones got on. This is the latest point in time when the Yugoslav escorts were got rid of. The journey continued. It was dark when the group arrived in the barracks on the outskirts of the city.

We can learn many interesting things from this 'story with a common denominator', especially things we would be glad to know more about. Why did the group decide to leave the embassy? What role did the Kádár–Münnich government and the Russians play in the abduction? What did the Yugoslavs know about that whole thing? What did they intend to do with the group? Did they influence Nagy and the others in their decision to leave the embassy? But problems and hindrances arise when we try to answer these questions on the basis of the narratives.

In our 'story with a common denominator' there are nine well-defined key groups of themes. But, within each group, the narratives contain many variations incompatible with each other, a phenomenon which will now be illustrated.

What was the motive behind the decision to leave the embassy? Among the explanations given were: political guarantees and the situation at the embassy—lack of space and food. *What were the causes of uncertainty before getting on the bus*? The Slovenian housekeeper burst into tears, saying 'stay here, don't go away'; the Slovenian housekeeper kept repeating to us in German in a desperate voice that we shouldn't get on the bus because the driver was a Russian; the cook working for the embassy, who came from Wojwodina, warned us in Hungarian that we shouldn't get on; a Hungarian official ordered us to get on; a Hungarian policeman arrived to say that Münnich[2] wanted to talk to the comrades. *What aroused suspicion while getting on the bus?* We already knew the Russian bus driver as he had been watching the embassy for a few days; when the passengers got on, Mrs Sz. said something to the driver who didn't understand a word of it. He had a typical Russian face; a Hungarian policeman was standing beside the bus and that was when he mentioned Münnich's summons; a Hungarian policeman from the state security forces, in civilian clothes, conveyed Münnich's request; two men in uniform were sitting in the bus; a Hungarian soldier with a machine gun was standing in the bus. *What were the reactions of people on the bus?* People already on the bus got off. They waited in the garden of the embassy, or they went back into the building. Nobody got on because Imre Nagy wanted to investigate the situation—why the driver was a Russian and what Münnich had to do with the whole thing?—while the people waited inside the embassy building.

What about the conversation between Nagy and Soldatić? It took place inside the building, or in the garden. It took place before they got off the bus, or after they had stopped getting on. *What languages did Nagy and Soldatić speak in?* Russian and French, and this led to difficulties in understanding because Nagy spoke only a little French and the ambassador only a little Russian. Or German, or Hungarian: 'I heard them with my own ears'. *What did the ambassador say? Among the versions:* Soldatić had already suspected something so he begged Nagy not to leave the embassy; after listening to Nagy he suggested that the group might remain in the embassy if he (Nagy) found the situation suspicious: he (Nagy) could remain in the embassy, for several years if necessary, while the women and the children could leave because he was sure nobody would harm them; Soldatić did not want to influence anybody so he did not participate in the decision in any way. *What about the decision to leave the embassy?* Some accounts state that after the conversation between Soldatić and Nagy the men discussed the situation and decided to leave the embassy; 'We made the decision together, all of us.' Others insist that Nagy made the decision alone, based on political considerations; or because the situation at the embassy had become very bad; or he waved in resignation and said nothing mattered, there was no point in changing the decision.

What of the bus itself? The windows were curtained so it was impossible to look out; or the curtains were drawn but it was possible to draw them back; there were no curtains. *Who else was there in the bus apart from the group and the driver?* A Hungarian soldier with a machine gun; men of unknown nationalities wearing police uniforms; a Hungarian policeman; a Hungarian civilian and a Russian officer; two men in uniform. *What of the Yugoslav escorts?* Two Yugoslav diplomats got on with us as observers; Yugoslav journalists got on with us as observers; members of the embassy staff accompanied us in two cars. *What happened to them later?* The two armed men of unknown nationality roughly pushed them off the bus; the Russian officer ordered them to get off the bus and when they refused to do so some Russian soldiers arrived and pushed them off; two Russian tanks held up the two embassy cars accompanying us. *Why did the bus stop on the journey?* It suddenly stopped so that the Yugoslavs could be pushed off; it stopped in front of the Soviet military headquarters' building, because the two accompanying Russian officers wanted to report that the action had been completed.

I suppose historians familiar with oral history will not be surprised by the inconsistency of the oral accounts describing such important historical events as those illustrated. This phenomenon is well known. However, the following questions cannot be evaded. How are the inconsistencies in these memories to be explained? Can we consider these oral accounts as historical sources at all, given these inconsistencies?

Our example undoubtedly proves that the 'unreliability' of the oral narratives cannot be attributed to forgetfulness. If we compare the 'story with a common denominator' with its very many variants, we can see that in fact the 'common denominator' is structural but is not valid in respect of content. Recollections have not shrunk to the 'essence of the matter' but have retained their individual differences right up until today. What is more, since some of the variations of the story are so rich, we could easily apply the notion of an autobiographical memory that has been imprinted on somebody's mind especially strongly. According to Linton[3] such memories occur under the following conditions: 'the event must be salient and be perceived as strongly emotional at the time it occurs, or it must be "rewritten" shortly thereafter; the life's subsequent course must make the target-event focal in recall—the event must be seen as a turning-point . . . the event must remain relatively unique. Its image must not be blurred by subsequent occurrences of similar events.'

The experiences of Imre Nagy's group undoubtedly meet all these conditions completely. On the other hand, as Neisser[4] has shown on the basis of analysing what he calls 'snapshot-experiences', the vividness and minute detail of memory does not tell us anything about the truth of the content. As Linton has already highlighted, the importance and central significance of the theme of recollection can also be created during 'rewriting' at a later time: the event is reconstructed both from the memory of realistic elements and from others made up later. Our example is a typical illustration of the process: the interviewees do not simply remember the common core of the story but richly build it up, so that at times one account differs widely in content from another, but is very similar to it structurally, even though specific vivid details sometimes plainly contradict one another. The shared experience is expressed through the similarity of the structure of the stories: the concrete elements of memory are arranged around the same dramatically significant turning-points. Thus, recollection has two levels: a content-concrete level and a structural-organizational one. The interviewees build up a narrative structure; they remember certain general features of the independent structural units, and finally they express these general characteristics of the situation through concrete episodes. When telling the whole story, they follow certain general rules of the narrative. They arrange the story elements into a context of place and time, and cast the people into leading and supporting roles; but the structure will emerge only later. It takes historical distance to decide clearly what the turning-points 'of a long-term determining effect' and the dramatically significant turning-points in the events actually were. By thinking it over several times, or by telling the story repeatedly, they will be imprinted on the memory. This is why the individual narrative structures are relatively similar—at the climax of the story even the individual episodes are identical. In our case nearly all interviewees remembered the moment when it became clear to them that they were prisoners, and they also remembered Mrs D.'s remark.

The general characteristics of the individual structural units are almost identical for each narrator. For example, there is consensus about the fact that the situation at the embassy was intolerable and that they were afraid of falling into a trap; about the unreliable behaviour of the Yugoslavs; and about the feeling that the decision to leave was a leap into the dark. But these general characteristics are no more than 'background content'. They are not specifically stated but are expressed only through individual variants of a great number of episodes, many of which have obviously been constructed much later. The general characteristics of the structural units are the cumulative result of a diffuse contextual knowledge—for example about the preceding events or the main characters (who most probably never said what they are quoted as having said, although they might well have said those things)—and the general coherence of the situation, which is often only later, after the event, reconstructed by generalization from concrete information and experiences which have already faded in the mind. For example, we get a more or less exact picture of the atmosphere of the discussion between Nagy and Soldatić and find out approximately what they talked about. But the narratives do not tell us what the two men actually said to each other, nor even what language they communicated in.

In psychological literature, this phenomenon has been the subject of controversy for a long time. Bartlett[5] emphasizes that even perception works in a selective way and that there is an overwhelming tendency to simply receive a general impression of the whole, on the basis of which the probable detail is constructed. 'When a subject is being asked to remember, very often the first thing that emerges is something in the nature of an attitude. The recall is then a construction made largely on the basis of this attitude, and its general effect is that of a justification of the attitude.' On the other hand Neisser,[6] analysing John Dean's evidence given at the Watergate trial, showed that truth appears in a narrative even if the individually-related stories do not describe the actual event at all. Even in this case the episodes can express a general semantic knowledge, and can typify or represent a line of events which have actually happened. Therefore Neisser believes that we should call this type of recollection 'repisodic' rather than 'episodic'.

The '(r)episodes' themselves, as well as the structure of '(r)episodes', are very important historical sources. Not only can they serve as the starting-point of semantic knowledge, suitable for reconstructing the diffuse background content, but the individual variants also give a lot of information about the narrators themselves. The '(r)episode' structure conveying general features often emerges under the constraint of questioning: the interview situation compels the interviewee to fill in his semantic knowledge with '(r)episodic' elements. This is one of the reasons for the emergence of individual variations. The selective construction of the episodes depends on several different factors, such as where the person's place was in the hierarchy of the group, the sexual role division within the group, the fears, expectations, and desires of the interviewees, and so on.

Differences in the situations they experienced and in the personalities of the observers most probably have an impact even at the stage of perception. The same differences will appear while 'coding' the perceptions. They are determined from the beginning by the knowledge that the person already possesses, for example about the character or the mentality of the protagonists. Finally, as Neisser and Nigro[7] show, the process of recollection can also be influenced by several other factors—the emotional weight of the event recalled, the length of time which has passed since the event, the motivation for recollection, and so on.

Can the oral narratives given by people participating in highly significant historical events thus be considered as historical sources? If we want to reconstruct the actual chain of events from the episodes as they are told, the answer is 'certainly not'. This has nothing to do with the specific memory of the interviewees but is a consequence of the special way in which memory operates in these circumstances. But if we both analyse the structure of the narratives and also remain on the level of semantic recollection—reconstructing the general characteristic features of the structural units and the elements of diffuse knowledge conveyed by the structure of '(r)episodes'—we might gain valuable historical information from these narratives, above all about the people who actively participated in the historical event, and influenced the outcome of events by their decisions and actions—just as in *Rashomon*, Kurosawa's film which is an outstanding example of his historical mode, we see and hear breathtaking variants of the same story, which help us understand—not so much the truth, but the meaning of a historical event, at a more profound level.

Notes

1. I have used 10 interviews for this analysis; 6 of them were recorded by Judit Ember in 1988 for her documentary on the abduction of Imre Nagy and his group—she recorded 9 interviews with participants in the same events.
2. The 'strong man' of the Kádár government.
3. M. Linton, 'Transformations of memory in everyday life', in U. Neisser (ed.) *Memory Observed: Remembering in Natural Contexts*, San Francisco, 1982, 85–90.
4. Neisser, 'Snapshots or benchmarks', in id., *Memory Observed*, 43–8.
5. F. C. Bartlett, *Remembering*, Cambridge, 1977, 206–7.
6. Neisser, 'John Dean's memory: A case study', in id., *Memory Observed*, 139–63.
7. Id. and G. Nigro, 'Point of view in personal memories', *Cognitive Psychology*, 15/4 (1983), 467–83.

8

Mujeres Libres

The Preservation of Memory under the Politics of Repression in Spain

MARTHA ACKELSBERG

As a contemporary North American feminist and political scientist travelling in Spain between 1979 and 1982, I confronted repeatedly, though in often unexpected ways, some of the more remarkable legacies of the Franco era: not only the virtual abolition of certain types of political memory, but also a dramatic disjunction between the perceptions (and self-perceptions) of those who were active in the social revolutionary movements of the 1930s and those of contemporary political activists. The phenomenon was particularly strong in the case of two groups of women who acted under the name Mujeres Libres: the *veteranas* (old ones, or veterans) who had been militants in the organization of that name, founded by anarchist women in 1936, and active until the end of the civil war, and the *jóvenes* (the young ones) who, whether in an act of solidarity or in an effort to endow themselves with a political past, took the name of the organization when they began to mobilize themselves into feminist groups after the death of Franco. Despite what seemed to me, at least, wonderful possibilities for political engagement across the generational barriers, and for translating the lessons of history into the present, the two groups proved—with a few minor exceptions—unable to work together and, beyond that, virtually unable even to talk without degenerating into mutual recriminations and misunderstandings.

I was struck by the extent and nature of the mistrust (and what I later came to see as misperceptions) on both sides. Typical comments from the *jóvenes* for example, included: 'you won't understand her (or them)'; 'they have crazy ideas'; 'they're not really feminists'; 'you'll be disappointed.' At best, they characterized the joint meetings as

The first version of this paper was prepared for the conference 'Spain 1936–1986', Center for European Studies, Harvard University, 14–16 Nov. 1986.

interesting (perhaps even enjoyable) events, but not as events which seemed likely to generate mutually-rewarding activities. From the *veteranas* the responses were somewhat different, though equally indicative of bewilderment: 'why do they spend so much time and attention worrying about issues like abortion, sexuality, or "love" '; 'they have no understanding of the politics of a movement'; 'they are too focused on "personal" issues'; 'how can they call themselves feminists (even libertarians!) and go around wearing crosses?'[1]

These comments reveal, I came to realize, both misunderstanding and disappointment on both sides. The *veteranas* had hoped to find a new generation to carry on their work and commitments; the *jóvenes* had hoped to find foremothers they could claim for their new political consciousness. Both groups felt cheated. To be sure, some people in each group did manage to sustain efforts at dialogue, and some are committed to continuing these. But the overwhelming sense was one of gap, of disjuncture.

What, I wondered, might account for such a rupture? Why could not 'feminist solidarity' make a sufficient bridge across the generations? Why should it be so difficult for two groups of people, who obviously had much in common and many similar goals, to share experiences and to build on one another's strengths? The answers to these questions are not simple; the process of transmission of political vision and political experience across time is a complicated one. But I believe that much can be learned from an examination of this phenomenon—in particular, from an exploration of what happened to the politics and the political lives of the women of Mujeres Libres during the Franco years—in enabling us to understand the nature and the costs of the transitions from social revolution to counter-revolutionary repression and then to democracy. The experience of these women has a good deal to teach us about memory and meaning; politicization, de-politicization, and politics.

Before we turn to an analysis of these questions, it might be helpful to review, briefly, the origins, ideology, and programme of Mujeres Libres, so that we have an accurate basis for exploring the changes these women underwent.

In the years leading up to the outbreak of the civil war, the institutions and practices of the Spanish anarcho-syndicalist movement had come to reflect its commitment to achieving an egalitarian, non-hierarchically-structured, communally-oriented society—and to doing so in a manner consistent with those ends. In the view of anarchists (and anarcho-syndicalists), a revolutionary process structures the society it aims to create. Therefore, the only way to create an egalitarian society, in which all are respected and treated equally, is through institutions and practices which are, themselves, egalitarian, non-hierarchical, and empowering, and which enable the oppressed to gain the experiences and self-confidence necessary to take their full places in a revolutionary movement to create the new society. People would learn how to conduct themselves in a future society through participating in activities which create what have recently been termed 'egalitarian social spaces'.[2] Hence the importance of decentralized, locally-based and directed, direct-action strategies, rooted in the experiences of people's daily lives, co-ordinated by loose federations, and organized by the power of positive example.[3]

This perspective on the process of revolutionary change had important implications for the involvement of women in the movement. As early as 1872 (at the Zaragoza conference of the Spanish Regional Federation of the First International) the organization committed itself to a goal of equality for women, and the full integration of women, as equals, into the paid labour force—a commitment restated and rearticulated in the early years of the twentieth century in the form of campaigns to organize women into unions and to struggle for equal pay for equal work. Nevertheless, it should be noted that this commitment was not always translated into practice: Many anarchists denied that women faced any special subordination as women, or that special attention was needed to address the situation of women. Furthermore, most women who did work for wages did so in their own or others' homes—either as domestics or as piece-work 'outworkers'. Catholic unions paid more attention to these latter groups than did the CNT (Confederación National del Trabajo) which, in practice (though not in theory), tended to treat women workers as threats to male wages, rather than as especially exploited workers (a pattern, of course, common elsewhere in Europe and in the USA). Still, among those women who were employed in factory jobs (primarily in textiles, in Catalonia), levels of unionization were high, and women participated in strikes at rates equal to, if not greater than, those of men.[4]

In addition to factory-based union organization, of course, movement organizations also sponsored and co-ordinated community-based struggle, and in these activities women consistently took an active role. Thus, for example, whether in the village-wide strikes in rural

Andalusia in the latter part of the nineteenth century, described by Temma Kaplan, or in the city-based bread riots, peace demonstrations, or 'quality of life' actions in Barcelona in the early years of the twentieth century, women's participation was formidable. What we find in examining such struggles is militant participation by women, some of it even crossing class lines, in which the CNT took a role, but did not necessarily lead.[5]

Finally, some women—though clearly not as many women as men—participated in and attended rationalist schools and *ateneos* (clubs), and took an active role in anarchist youth organizations, in the years before the civil war. Thus, when the war began, many women had had experiences of active participation in the movement; and the movement, in its many facets, had manifested a commitment, albeit with varying degrees of seriousness, to overcoming women's subordination. The way had been cleared for widespread participation by women in revolutionary activity.

During the civil war itself, women participated alongside men in the popular response to the generals' rebellion, and in the social revolution which took off from it. In cities such as Barcelona and Madrid, women joined actively in the initial anti-fascist mobilization, constructing barricades, storming the armouries to get weapons, joining the militias, providing first aid to militiamen (and women), and setting up popular kitchens. As one women reported,

> They sent me, and all the women and families, to building barricades. And then, too, we took care of food—all the women in each *barrio* organized that, to make sure, especially, that there would be food for the men. Women participated a lot in those days—especially the first few days. Everyone did something—whether it was building barricades, organizing food, sewing clothes, socks, or whatever.[6]

Women also engaged in the process of revolutionary collectivization, both industrial and rural. In urban-industrial areas, the war and revolution resulted in a massive movement of women into the paid industrial work-force, particularly in the new chemical and metallurgical industries. In addition, they replaced men in areas such as public transport—driving buses, trams, and taxis for the first time. The textile industry was reorganized by the CNT to eliminate piece-work and outwork, in an attempt to improve working conditions for the majority of women. Similar efforts were made with respect to beauty parlours, barber-shops, bakeries, and other small workplaces: small, unhealthy shops were closed down, and workers moved to larger, more airy, and better-lit locations. In other industries (most notably wood, for example), unions constructed day-care centres and schools on the premises, to facilitate the full incorporation of women into the paid labour force. In agricultural areas, women's participation and pay varied: sometimes they received wages equal to those of men; sometimes they were incorporated into a 'family wage' system. But virtually everywhere, women experienced considerably greater degrees of social freedom, and many more opportunities for informal male–female interaction.[7]

There were, of course, limits to the achievements—and particularly to the degree of equality for women. Equality between men and women was, in fact, rarely a goal for the collectives, and true wage equity was a rarity. Neither men nor women in the movement evinced much awareness of the ways in which the sexual division of labour, and differing expectations for women in the home, would affect women's participation at work, whether as workers or as union or factory delegates.

Meanwhile some of the women who had joined the organizations of the anarcho-syndicalist movement had come to recognize limits to what they could achieve within the structures of the CNT. On their view, women were not always treated with the seriousness and respect they deserved: much more than men, for example, women tended to be ridiculed for their ignorance. And they were all too aware of the inability of the CNT to attract many competent women to its ranks—to keep them coming after a first meeting or to move them into positions of leadership in unions. They attributed this failure both to the sexism of the men in the CNT and to the 'lack of preparation' on the part of the women.

As early as 1934, some of these women (notably Mercedes Comaposada and Lucía Sánchez Saornil in Madrid) began talking, sending out letters, and writing columns in *Tierra y Libertad*, raising the issues of women's subordination and the need to address it within the CNT. Eventually, these letters and conversations led to the formation of Mujeres Libres (in Madrid) and the Grupo Cultural Femenino (in Barcelona), soon incorporated into a broad federation under the name Mujeres Libres, and the establishment of a journal under the same name in May, 1936. The organization had as its goal the *capacitación* (empowering) of women to take a fuller part in the libertarian movement: 'The emancipation of women from their triple servitude: enslavement to ignorance, their enslavement as producers, and their enslavement as women.' They argued that women had to organize independently of men, both to overcome their own subordination (to develop self-confidence and a sense of their own competencies) and to struggle within the organization against male resistance to women's emancipation.[8]

Mujeres Libres' programme was based on the same commitments to direct action and equality which informed the broader anarchist movement. Thus, their primary focus was on overcoming ignorance and illiteracy, which they saw as the most significant barrier to women's active participation and self-confidence. They sponsored literacy programmes at a variety of levels, both in the major cities and in rural areas, to enable women to gain the skills necessary for them to participate more fully in political and social life. They also developed programmes on health care, sexuality, maternity, and infant and child care, for all of which there was a considerable need and interest. Together with CNT unions, they mounted apprenticeship programmes at many factories (and in agricultural areas, as well), designed to enable women to move into responsible positions in industry, and to overcome the extreme sexual division of labour. Finally, they developed a variety of 'consciousness-raising' and empowerment programmes directed specifically at enabling women to take their full places in the revolutionary movement: 'flying day-care centres' which sent baby-sitters to the homes of young mothers, so that they could leave to attend union meetings; special meetings of women at the workplace in which they could speak among themselves, develop their ideas, and overcome their reticence to speak in public—especially in front of men.

Although their achievements may have been limited by the dislocations of war and by resistance within the movement itself, participation in the activities and programmes of the anarcho-syndicalist movement in general, and in those of Mujeres Libres in particular, transformed the lives of the *veteranas* I interviewed. Never before had they lived their lives at such a 'pitch' of activity and commitment; never before had they felt so much a part of a world-changing process. It was an experience—despite the hardships which accompanied and followed it, which remained with them for life. Over and over again, I heard comments such as these:

the times that we lived during the war, six months were like three years in another context . . . So that, for me, the three years of war, all that I lived through, were like . . . ten years of my life . . . When I was fourteen and fifteen I had experiences that would stay with me all my life engraved in my mind, such a flowering of ideas-made-reality that happened during this period! Even if I had died, I wouldn't have wanted not to have had that experience. (Pepita Carpena)

Or, in the words of another,

It was an incredible life, the life of a young activist. A life dedicated to struggling, to learning, renewing society. It was characterized, almost, by a kind of effervescence, constant activity . . . It was a very busy life, working eight hours—or sometimes ten, if we got overtime—and going everywhere on foot to save money for the organization.

The world seemed to have changed, dramatically. In the words of Anna Delso, describing the sense of responsibility felt by those working in collectivized industry in Villanueva i la Geltru, outside Barcelona, 'Each one gave the best of him or herself. Everything depended on us alone, and we knew it. It was a matter of life and death . . .' A founder of the Mujeres Libres group in Villanueva, Delso described her experiencing of women's empowerment:

I was astounded by women's capacity for organization. Many of them played a major role in the CNT local, and at the same time were members of the self-management committee at their factory. They rediscovered themselves as equals in a non-hierarchical society. It was a total and radical transformation of social life. And did Spanish women ever need it! They had been weighed down by the burdens imposed on them by Church, husband, father, brothers, and all the rest. To all those who said to us, 'Yes, we are in agreement with your goals as women, but you must leave them till later so as not to cause divisions within the ranks', we replied, 'When later? It's immediately, or never.'[9]

The sense of community they experienced in those times—both among themselves as women, and with men in the movement—was powerful. Time and again, discussion of those years, or of their exploits, brought tears to their eyes, or a special glow to their faces. The Spanish revolution was probably the largest, most varied, social revolution in the world; and probably the most complete social revolution that had ever taken place in the context of a war. For some who lived through this period—those who felt themselves caught up in events beyond their control and not of their own choosing—of course, memories of the war years may well evoke only pain. But for these women who felt themselves fully engaged in the revolutionary project, memories were powerful and positive. As participants in it, they had experienced a depth of commitment and, a sharing both of vision and of hardship which are rare, and which they had no desire to forget.[10]

Nevertheless, experiences of such intensity are, of course, difficult to communicate to others, especially those separated from the events by time, political context, and politics. And in the case of the *veteranas*, the process of transmission was further complicated by the years of repression, exile, and fundamental breaking of connections which the Franco era represented.

While their stories differed, for each the end of the revolutionary period marked the beginning of a very different, and much longer, phase of their lives: for those who remained in Spain, years of repression and fear, for those who went into exile in France, concentration camps, hardship, poverty, and, slowly, new beginnings—of a phase they assumed would be temporary, but which lasted for longer than any had dreamed possible. Those years marked them deeply: most were cut off from their roots, many from their families. The 'highs' of the revolution were matched by the 'lows' of the concentration camps and the deprivation which followed.[11] Remarkably, though, few lost their faith in humanity or in the value of the ideals for which they had fought.

The grounds for disillusionment, of course, were many. Those who conceived of themselves as the first fighters against world fascism, and who had already felt betrayed by the Allied policy of 'non-intervention', expected to be met as heroes when they crossed the border into France. Instead, they were treated effectively as criminals (when they were allowed in at all!), disarmed, and thrown immediately into makeshift camps on the French Mediterranean coast: Argèles-sur-mer, St Cyprien, Barcarès. With inadequate nourishment, shelter, or means of keeping warm, many spent the months of January, February, and March 1939 wondering whether they had escaped Franco only to die in France. Stein's summary of refugees' reports matches well those I heard: 'Spanish refugees are unanimous in their recollections of the mind-deadening and body-destroying bleakness of their early days in the camps.'[12] Many spent the days, weeks, and months wandering through the camps in hopes of finding, or hearing news of, brothers, sisters, parents, or friends. The feelings of demoralization and abandonment were extreme.

Once they had made it out of the camps—some by escaping, others released to work 'farmland' which had been abandoned as hopelessly infertile by its French former tenants—their troubles were far from over. The fear of being rounded up or turned over to the Spanish police or, after the occupation of France by Germany, to the Gestapo was ever-present. One women described living in one of these abandoned farmhouses for years, with no furniture other than a few crates, not wanting to accumulate possessions that would impede their eventual return to a free Spain. But of course, the 'free Spain' of their dreams was not to be—not for many years. For, at the conclusion of the war, when Spaniards both in Spain and in France expected the victorious Allied troops to march down over the Pyrenees and 'finish the war' by liberating Spain, they were disillusioned and abandoned once again. The Allies undertook no southern campaign; Franco remained in power, only to increase the repression once more.[13]

After the war ended, many—if not most—of the exiles remained in France, but with a firm identity as *refugiados*, resisting full integration into French society (even if it had been an option open to them), expecting, at any moment, to be able to pack up home and family and return to Spain. To live in such a manner, as a temporary sojourner—at best—in a land not one's own, is to live a life of perpetual rootlessness and isolation. Some refugees found their way to towns with former friends, comrades, or family (the Toulouse area, Béziers, Paris, for example), and managed to re-establish a sense of community and belonging, at however great a distance from home. Many did not, however, and lived a life—whether elsewhere in France, England, Canada, the USA, or Latin America—of considerable isolation and loneliness, cut off from family, friends, and—importantly—the political community which had given meaning to their lives.

Over time, of course, it became increasingly clear that they would not return to Spain, at least not in the way they had hoped. Children were born, and many married—some to children of other refugees, but many to natives of the countries in which they were living—and grandchildren often followed. Despite themselves, many of the refugees found themselves permanently caught in the lands of their exile: dependent on their work or—once Franco died, and the possibility of return to Spain opened up again—more likely, on government retirement pensions for their survival, with little or no family in Spain that they could turn to, and with children firmly rooted in their new country, unwilling or unable to return to Spain. Many, therefore, have remained in exile, despite the death of Franco and the 'liberation' they had so long awaited.

A few, on the other hand, have returned to Spain. Some did so even while Franco was alive; many others moved back only after he died. But these, too, now live very differently from the way they did in the 1930s. Even those who settled in major urban centres such as Barcelona, for example, who managed to re-establish ties with old and new comrades, and who attended meetings of a revived CNT, noted important differences. The political climate is not what it was. The movement no longer has the power and the presence it once did. As retirees, they cannot engage in working-class struggles in the same way they did as youths. The ties remain; and the power of the example of the past. But they live in a very different world—whether it be inside or outside Spain.

One further aspect of the experience of exile and repression merits attention here: the transmission of values across generations within families. While I do not have sufficient data to make a conclusive case, the patterns seem revealing. I noted in an earlier study that, in the absence of a continuing connection to a sustaining community, many former participants in collectives seemed to lack any framework for analyzing their experiences, and any sense of the significance of what they had done. In the most extreme cases, even for some who had been activists, their experiences in the revolutionary period became a kind of 'unmentionable history'—events they could not discuss even with neighbours or children, sometimes lost even to themselves.[14]

Nevertheless, despite the disillusionment and the years of exile, this did not seem to be the case with any of the women of Mujeres Libres whom I interviewed. One after another told me proudly of the political commitments and activism of their children—few of them, of course, anarchists, since there is no anarchist movement to speak of in France; but all of them active on the left. One reported proudly that all three of her daughters had records with the French police: one for the May 1968 events; and two for helping refugees illegally over the French–Spanish border. Somehow, although they lived apart from any meaningful community of comrades, a life far different from the 'continual effervescence' they had experienced during the war and revolution, they had managed to communicate the power of their commitments to their children, who had, in turn, made them their own.

I found no similarly consistent pattern of 'successful transmission' of values among the men I interviewed, though I am not sure what to attribute the difference to. Perhaps because women remained the primary child-raisers, even among committed anarchists, they were able to communicate their values in a very direct fashion to their children. Perhaps it is the fact that, while most women activists seem to have married—or set up house with—other committed CNT activists, this was not true of all the men: thus, those children growing up in the homes of the *veteranas* may have had greater, and more consistent, exposure to leftist ideas than those who grew up in the homes of male activists. Whatever the causes, the difference seemed worth noting.

The *veteranas*, then, had lived a life of great intensity, both during the period of the revolution and during at least the first few years of the repression which followed. More significant than the intensity itself, however, is the content of it—and the context within which they lived. A number of factors seem particularly important, both in characterizing their lives and in differentiating them from those of the *jóvenes* who would try to understand them. First, the importance to the *veteranas* of the *movement* as a context for activity and for meaning; second, the impact on them of the national and international *political context* in which they were operating, both during and after the revolutionary period; and, third, the nature of their understandings of '*feminism*'.

First, community. It was clear that, for all the *veteranas* I interviewed (as was the case with the male activists too), the movement—whether the Juventudes (the anarchist youth organization), the CNT, Mujeres Libres, or some combination of them—had provided their primary source of identification, the shared community of meaning which had given order and direction to their lives. Pepita Carpena, for example, referred to the local metalworkers union, whose meetings she had started attending at age thirteen, as her *sindicato de corazón* (the union of her heart), even though she worked as a seamstress and had organized a seamstresses' local at her own place of work. Nevertheless, it was the metalworkers who had made a place for her, who had won her heart. It was they who supported her with a weekly stipend throughout the war as she engaged in travels and propaganda trips for Mujeres Libres.

Enriqueta Rovira spoke similarly of her and her sisters' experience in their local *ateneo*, Sol y Vida: 'The group was something very wonderful, very special. That's where ties between people really formed . . . That's where our ideological development really happened.'[15] The bonds among *compañeros*, the sense that one had a community of people to count on, was, for all of those I interviewed, a central feature of the movement. As Azucena Barba, one of Enriqueta's sisters, put it:

'Our house was always open. There is a certain Spanish "hospitality", in general, which tends to be warm and open to people . . . But that kind of hospitality was a *special* part of our ideas, of our movement. No matter how crowded we were, there was always room for another. That, yes, that is and was a very important part of what we were up to.'[16] In short, even among those who had come from homes where their parents were anarchists, their primary community, meaning, and identification was not the family, but the movement. That was obviously all the more the case for those whose political commitments set them apart from more conservative, or apolitical, parents.

Secondly, the political context. The sense of abandonment by supposedly friendly political forces outside Spain, which engendered a combination of anger and bewilderment, was common for all those in the republican camp, including conservatives, liberals, socialists, and communists, as well as anarchists; but it was particularly acute for anarchists. For, while the Soviet Union (and Mexico) eventually came to the aid of the republic during the war, the price of USSR support was the strengthening of the Spanish Communist Party and the PSUC (Partido Socialista Unificado de Cataluña) in Catalonia, and the suppression of revolutionary movements within the republican camp. That process began most dramatically with the 'May Days' in Barcelona in 1937—which resulted in the suppression of the POUM (Partido Obrero de Unificación Marxista) and the CNT—continued with the march of Lister's column through Aragon to disestablish rural collectives in August of that year, and resulted, as well, in moves to limit the autonomy of industrial collectives. Even while the war and revolution were being fought, then, members of anarcho-syndicalist (and left-socialist) groups felt themselves embattled and isolated. They could turn to no one but themselves for support—a phenomenon which, no doubt, also increased the importance of community to them.

Once the war ended, and the years of exile or hiding began, the sense of isolation only increased. There was an international communist movement (and the Soviet Union) to take in the leadership of the Communist Party, and to nurture their underground organization during the years of Francoist repression; and European socialist parties to support those who were to become the new leaders of the PSOE (Partido Socialista Obrero Español). But there was no comparable European anarchist or anarcho-syndicalist movement on which the members of Mujeres Libres could rely. While some members of the CNT and of Mujeres Libres maintained ties with one another and with other similarly-inclined Europeans through organizations such as Solidaridad Internacional Antifascista, most found themselves isolated politically as well as socially in their new homes. That is not to say, of course, that they stopped acting politically: they fought in the French resistance, joined labour unions at their places of work, and formed coalitions where they could. But they never recovered the kind of connection which had been so definitive of their lives during the war and revolution. Although many were able to hold on to their own beliefs despite this isolation, their feeling of being cut loose from their moorings remained powerful.

Thirdly, it is important to be aware of the consequences of their political context for the *veteranas'* understandings of feminism. The women of Mujeres Libres, though committed to the emancipation of women, did not define themselves as feminists. In fact, for many of them, the word was anathema: they agreed with Emma Goldman and Federica Montseny that feminism referred, essentially, to a bourgeois women's movement, a demand for voting rights and equal professional opportunities, neither of which seemed a valued goal to anarchists! From their point of view, the crucial issues were economic—class-based—and the struggle for women's emancipation could only be properly understood and accomplished from that perspective. Strategies or conceptions of women's emancipation which ignored the concerns of working-class women, or more accurately, failed to put them at the centre were totally irrelevant.[17]

If we compare the life experiences of the *jóvenes*, we find that they differ significantly from the *veteranas'* in each of these areas. First, probably the most important legacy of Franco's repression—particularly for that generation which came of age in Spain during his rule—was the abolition of political memory. History, if it was taught at all, was taught as the victory of God and country over the godless hordes. Nothing of the revolutionary achievements of the civil war period was communicated to the young people who grew up in the post-war period. Nor was it possible to speak of, and barely possible even to read about, the ideals and values which had motivated the republican forces, let alone the anarchists. Thus, it was virtually impossible for young people within Spain to learn about what had happened, even by word of mouth from parents or grandparents. So, while they may have learned about political movements, even student protests, elsewhere in Europe, most were denied knowledge of—and, therefore, any understanding of the significance of—the events which had shaken

Spain. Of course, many attempted and found ways to learn about these events despite the dangers. They read books, or talked to participants, in secret; they jammed the bookstores and demonstrations in the first years after Franco died and *veteranas* and *veteranos* began to tell their stories once again. Nevertheless, they were, at the least, denied any experience of the kind of intense anarchist community which had given meaning to the early lives of the *veteranas*.

Second, and related to this, the *jóvenes* came of age, and now live and work in the absence of either a national or an international anarchist or anarcho-syndicalist movement. While the 1920s and 1930s were highly politicized—at least for those *veteranas* with whom we are comparing them—the *jóvenes* grew up in an apolitical world in which to talk politics was to risk one's life. The political affiliations which are now 'available' to them are very different from those which were available to the *veteranas*—socialism, feminism, anti-nuclearism—and while they obviously share goals and perspectives with anarcho-syndicalism, they are different both in scope and in method. Finally, the overall political and social climate has changed dramatically. Some of the central problems which confronted the republican government of 1931–6, and which formed the basis for the political programme of the CNT and Mujeres Libres, have, effectively, been solved. To mention just two, the so-called 'modernization of agriculture' which resulted in the virtual depopulation of many ares of Andalusia and Extremadura, for example, has rendered moot one of the major issues which confronted the republic and provided militants for the CNT in the early years of this century, namely, the redistribution of agricultural land. Industrial modernization has provided more jobs, and allowed for working conditions which are vastly improved compared to those which prevailed during the pre-civil war period.[18] In short, not only do they lack the experience of participating in an anarchist political–social community; the political context in which they have grown up and in which they now act is vastly different from that in which the *veteranas* moved.

Finally, and related of course to what has already been said, feminism has come of age in the interim. While the *jóvenes* have not had an anarcho-syndicalist movement to relate to, there is a growing feminist movement, both in Spain and internationally. That movement has focused not simply on the right to vote, and on professional equality—which characterized the early feminist movement in Spain—but, more importantly, on abortion, divorce rights, equal pay, and educational equity. All of these, in some degree, are manifestations of a central feminist tenet that 'the personal is political'. Contemporary feminists both in the USA and western Europe, and, to a considerable degree in Spain, reject the distinction between political and personal concerns which characterized much of the traditional left—as well as much of traditional politics—and they insist on the political nature of personal issues. For contemporary feminists, including many of the *jóvenes* active in Spain in the late 1970s and early 1980s, issues of sexuality, love, abortion, and divorce are as political as workplace struggles—if not more so. In addition, many of the *jóvenes* came to their feminism at least in part through disillusionment with the ways they were treated by mainstream political organizations, particularly the socialist or communist parties.[19] Their 'community of identification', to the extent that they had one at all, tended to be other feminists rather than other people of their movement—again marking a significant departure from the experiences of the *veteranas*, whose community of orientation was not other feminists, not even other women, but other anarchists.

What, then, has been the legacy of the activism of the women of Mujeres Libres? And what can we learn from it about the creation and transmission of political meaning during a period of political repression?

For the *veteranas*, the most important legacy is the sense of their own empowerment, of having lived and acted in such a way as to make a significant difference in the world. That, it seems to me, is what gives power to their reminiscences of revolutionary activism and to their memories of walking through Barcelona and feeling 'that the whole town was ours'. Such moments are rare—particularly in the lives of poor and working-class women who were born in an era when nothing more was expected of them than to engage in hard physical labour from dawn to dusk to support and maintain their families. For the women of Mujeres Libres, and for many other anarcho-syndicalist activists I've interviewed, that sense of empowerment was intimately bound up with their feelings of connection to the larger anarcho-syndicalist movement, and to the cause for which they jointly fought. Yet those feelings of empowerment and connection were tempered, in the post-war period, by feelings of betrayal and abandonment with respect to the so-called democratic forces of western Europe.

For most of these women, then, the legacy of their activism is a profoundly ambiguous one. On the one hand, they have powerful memories of their own actions in the course of the revolution to change, fundamentally, the nature of life and politics in Spain at the time. On the other hand, primarily because of the repression, their years of exile, and the markedly changed political climate in Spain and in Europe more generally, they have effectively been denied progeny. They can speak with the *jóvenes* of this generation only with difficulty, as they must cross significant boundaries of politics, community, and meaning in order to do so. While some have committed themselves to this effort, and have managed to bridge at least some of those gaps with grace and humour,[20] for many the task is significantly daunting. They are, to be sure, eager to speak and to communicate about their experiences. But sometimes it seems virtually impossible to surmount the real cultural and political differences which separate them from the *jóvenes*.

For the *jóvenes*, the legacy of Mujeres Libres' activism has been all but lost. Occasionally, articles appear in the popular or feminist press,[21] telling the stories in an attempt to reclaim a history. Books by Mary Nash, Lola Iturbe, Lidia Falcón, and others, and the recent 'Homenaje a las Mujeres de la Guerra Civil' held in Barcelona in the fall of 1987 make the experiences and activities of Mujeres Libres, in particular, and women of the anarchist movement, in general, available to broader audiences. But, again, largely because of the forty years of repression, and the markedly changed political context today, they have found relatively little way to connect politically with the *veteranas*. The desire for connection is there; but the means to overcome the significant differences dividing them have not yet been found.

These observations suggest to me that the situation of the women of Mujeres Libres may provide a particularly stark example of what happens to the legacy of revolutionaries in a period of significant political repression. While the revolutionaries themselves may be able to hold on to the image and power of what they had accomplished, it is virtually impossible for them to pass that knowledge on to others, particularly to others separated from them in time and space. The combination of lack of access to the people who participated in these activities, and significant censorship of materials which might have told the story of their activities, made it virtually impossible for those coming of age under Franco to develop a sense of the nature of the revolutionary era through which they could relate to the *veteranas* once the 'opening' came with Franco's death.

In turn, this overview of the politics of the *veteranas* in comparison with that of the *jóvenes* also suggests some reasons for the apparent ease of the 'transition to democracy'. In the past—and in particular, in 1931–6—attempts at democratic transitions were foiled by a combination of oppositional forces on both the left and the right. But what Franco's repression seems to have accomplished—both through the jailing and execution of political dissidents and also hence the ability of the regime to force through various economic 'modernization' programmes without fear of opposition from workers—at least in the case of the women of the anarcho-syndicalist movement, and I suspect it could be generalized to the movement as a whole, is the effective destruction of that oppositional movement. Yet the absence of the institutions and organizations of the movement makes it significantly more difficult to communicate, in understandable ways, even the history of what was accomplished fifty years earlier. The relative ease of the 'transition', that is, is assured by precisely the same factors that rob the *veteranas* of their progeny, and the *jóvenes* of any proper sense of their foremothers.

Notes

1. It took a while to realize what the issue was in this case; but the solution was discovered by Ana Cases—a *joven* who was committed to talking with *ancianas*—in a conversation with a *veterana* at a CNT congress in 1980. It seemed that what the older women had taken to be crucifixes were, in fact, women's symbols (♀), which a number of the younger women were wearing as signs of their feminism!
2. See, for example, Sara Evans and Harry Boyte, *Free Spaces: The Sources of Democratic Change in America*, New York, 1986.
3. For a fuller exploration of the theory and practice of 'preparation' see Martha Ackelsberg, 'Mujeres Libres. Individuality and community: Organizing women in the Spanish Civil War', *Radical America*, 18/4 (1984), 7–19; and ead., 'Contexts of revolution: Sex roles and anarchist collectivization in civil war Spain', in Yosef Gorni, Yaacov Oved, and Idit Paz (eds.), *Communal Life: An International Perspective*, Efal (Israel) and New Brunswick (NJ), 1987, 591–605.
4. See, for example, Rosa Maria Capel Martinez, *El Trabajo y la educacion de la mujer en España (1900–1930),* Madrid, n.d.; Alberi Balcells, 'Condicions labórals de l'obrera à la industria catalana', *Recerques*, 2 (1972), 141–59; and id., 'La mujer obrera en la industria catalana durante el primer cuarto del siglo XX', in *Trabajo industrial y organización obrera en la Cataluña contemporánea, 1900–1936*, Barcelona, 1974, 7–121.
5. On these struggles see, especially, Temma Kaplan, *Anarchists of Andalusia*, Princeton, 1978; and 'Female consciousness and collective action: The case of Barcelona 1910–1918', *Signs*, 7/3 (Spring 1982); Lester Golden, 'Les dones com avantguarda: El rebombori del pa del gener 1918', *L'avenç* (Dec. 1981), 45–52; and Ackelsberg and Myrna Breitbart, 'Terrains of protest: Striking city women', *Our Generation*, 19/1 (Fall 1987), 151–75.

6. Enriqueta Rovira in interview with author, Castellnaudary, France, 28 Dec. 1981.
7. For further details about the process of collectivization—especially its achievements and limitations for women—see Ackelsberg, 'Contexts of revolution'; and ead., ' "Separate and equal"? Mujeres Libres and anarchist strategy for women's emancipation', *Feminist Studies*, 11/1 (1985), 63–83.
8. For a fuller discussion of the founding and goals of the organization see id., 'Mujeres Libres. Individuality and community'; ead., 'Separate and equal'; Kaplan, 'Other scenarios: Women and Spanish anarchism', in Renata Bridenthal and Claudia Koonz (eds.), *Becoming Visible: Women in European History*, New York, 1973, 400–21; Mary Nash, *'Mujeres Libres' España, 1936–1939*, Barcelona, 1976; and *Mujer y movimiento obrero en España, 1931–1939*, Barcelona, 1981 (esp. ch. 2). My own work on Mujeres Libres forms a part of Ackelsberg, *Free Women of Spain: Anarchism and the Struggle for the Emancipation of Women*, Bloomington (Ind.), 1991.
9. Anna Delso de Miguel, 'Démarche évocatrice', in, *Trois cents hommes et moi ou estampe d'un révolution*, Montreal, 1989, 49.
10. I have explored the processes of politicization and depoliticization (and the consequences of the latter) in Ackelsberg, 'Revolution and community: Mobilization, de-politicization and perceptions of change in civil war Spain', in Susan C. Bourque and Donna Robinson Divine (eds.), *Women Living Change*, Philadelphia, 1985, 85–115.
11. For a powerful description of the deprivations the refugees suffered in France in the post-war period as well as their role in the French resistance, see Louis Stein, *Between Death and Exile: The Spanish Republicans in France, 1939–1955*, Cambridge (Mass.), 1979; Nancy Macdonald, *Homage to the Spanish Exiles*, New York, 1987, tells a similar story.
12. Stein, *Between Death and Exile*, 57.
13. Those who had remained in Spain, and taken part in underground resistance there, during World War II, reported a general expectation, in 1943–4, that the war would be carried over the Pyrenees and that Franco's regime would be ended. Arturo Parera, for example, who had organized miners in the north of Spain, reported that, as the war was winding down in France, employers were falling over themselves to pay 'dues' to the CNT, in hopes that, once the regime fell, they would be on good terms with the labour unions which they expected would return to a position of power. Meanwhile, jailings and repression within Spain had lessened somewhat, apparently in anticipation of the 'liberation', but the rate of arrests and prosecutions picked up again dramatically after the war ended in France. In addition to Stein, information taken from interviews with Arturo Parera, Sitges (Barcelona), 22 July 1979; and Valero Chiné, Fraga (Aragon), 11 May 1979.
14. See Ackelsberg, 'Revolution and community', 102–8.
15. Enriqueta Rovira in interview with author, Castellnaudary, France, 28 Dec. 1981.
16. Azucena Barba in interview, Perpignan, France, 1 Jan. 1982. I heard similar comments from Lola Iturbe, Sara Guillén, Mercedes Comaposada, and many others.
17. See, for example, Ackelsberg, 'Mujeres Libres. Individuality and community', esp. 14. See also Emma Goldman, 'Women's suffrage' and 'The tragedy of women's emancipation', in, *Anarchism and Other Essays*, New York, 1969, 195–212 and 213–26 respectively; and Federica Montseny, 'Feminismo y humanismo', *La Revista Blanca*, 33 (1 Oct. 1924), and 'La tragedia de la emancipación feminina', ibid. 38 (15 Dec. 1924). The issue was brought home to me repeatedly in interviews with Spanish anarchist women.
18. It is important to note, in this respect, that both of these phenomena were achieved with tremendous cost: rural poverty, migration of workers to the north of Spain and elsewhere in Europe, and severe repression of labour unrest. That is, many of the

'successes' of the Franco era were, in fact, made possible only by the repression which was its hallmark.

19. Note the parallel with the civil rights and anti-Vietnam War movements which provided an important context for the 'second wave' of feminism in the USA. On these roots, see Sara Evans, *Personal Politics: The Roots of Women's Liberation in the Civil Rights Movement and the New Left*, New York, 1980.
20. Pepita Carpena comes most clearly to mind. Despite her frustrations with the focus on 'the personal'—which, in her view, characterizes all too much of contemporary feminist rhetoric and activism—as director of the Marseilles office of CIRA (the Geneva-based International Centre for Research on Anarchism), she frequently attends conferences and other gatherings, and eagerly attempts to engage in dialogue with *jóvenes*.
21. e.g. *Vindicación Feminista, Historia de España, Cambio 16.*

9

A Shattered Silence

The Life Stories of Survivors of the Jewish Proletariat of Amsterdam

SELMA LEYDESDORFF

It is commonly assumed that eyewitness accounts are created by the interaction between the interviewer and the interviewee and that, given the problematic character of interviews as a historical source, the resulting historical information needs to be examined thoroughly. Yet in some cases they are the only source (although a very problematic one) available to the historian. They also provide us at times with a new kind of information about, for instance, the experience of a certain epoch. Such is the case for the researcher into the Jewish past who tries to unearth a history that has been distorted by mourning and veiled by the longing of the interviewed for bygone days in which the great evil had not yet arrived.[1]

In this case one very often has to make use of the tales of severely traumatized people, in the knowledge that their stories have been formed by the way stories of survivors have been used in the past. The stories relate again and again to other eyewitness accounts and to the massive interest generated by a specific and sometimes stereotypical narrative of the Holocaust. One example of this is the recent attention given to Klaus Barbie, who, had he not been tried with such tremendous publicity, might have remained one of the nameless torturers, only a little worse than other brutes.[2] Another problem is the fact that we are dealing with a layer of memory constructed fifty years ago and sometimes under extreme conditions. Several authors have

S. Leydesdorff, *Wij hebben als mens geleefd: het joodse proletariaat van amsterdam 1900–1940*, Amsterdam, 1987 (American translation forthcoming: 1992, Wayne State University Press).

A very short version of this article appeared as 'Das gebrochene Schweigen, Lebensgeschichten von Überlebenden des jüdischen Proletariats in Amsterdam', *Bios: Zeitschrift für Biographieforschung und Oral History*, 2 (1988), 17–27.

stressed the mythic character of the tales of the events in the camps themselves, and to one acquainted with the atmosphere surrounding the recent legal proceedings against war crimes it is clear how strong this myth-making impulse is. In his book about the misunderstandings between the insiders (camp inmates) and the outsiders (those who listened to them in court), the Belgian historian van den Berghe has illustrated the inmates' need to create certain stories. They sometimes did so in order to survive. He illustrates this with the various narratives surrounding the death of Mala Zimetbaum, who tried to escape with her lover Edward Galinski. Van den Berghe researched the many eyewitness accounts of the ways in which they were killed and says:

> Around these events and about the relationship between Mala and Edek real epics flourished immediately. These were interwoven with several themes that appeal strongly to the hearts of prisoners, survivors, and outsiders. The courage of those two young prisoners, their relationship, and their heroic death create and created an ideal point of crystallization for the hopes, self-criticism, and feelings of guilt of many. Functioning as a sign of hope or self-criticism for the prisoners, or as a way to give a rationale for what they had interpreted as 'cowardice', 'lack of resistance', or 'guilt', certain details were emphasized, added, or omitted. Their stories often contain a very loaded or moralizing message.[3]

It is not acceptable to express any doubts about the veracity of these stories, since they are a message to the world from those who have gone through experiences which normal language cannot describe.[4] Only in the form of testimonies, like those of Primo Levi[5] or Victor Fränkl,[6] or in the poetry of Paul Célan,[7] or perhaps in a film, such as Claude Lanzmann's *Shoah*, can aspects of this world be transmitted.

The eyewitness account has, from the beginning, been among the major sources for writing the history of the Holocaust. It was one of the main objectives of the Nazis to bury all traces of their crimes; and although historians have shown that, despite the Nazis' efforts, proofs of these crimes can sometimes be found in train schedules and other indirect sources,[8] the stories of the survivors remain our major source. These alone can give us an account of the day-to-day suffering, and of the strength of those who fought the process of depersonalization.[9] But we must still realize that they are *survivors*, and even they can only fantasize about the horrors of those who were directly herded from the trains into the gas chambers. In a way, they are also the witnessess of other people's suffering and give such testimony as we have of the fate of those who did not survive.[10]

Testimonies are often labelled as 'subjective' or 'biased' in the legal proceedings concerning war crimes. The lawyers of war criminals have asked the most impertinent questions of people trying to find words for a shattered memory that did not fit into any language, or who could not find sentences to describe those days and months in which the only chance of survival was to forget that there had ever been a world of goodness, warmth, and beauty.

The conduct of these war-crime trials has led at times to intolerable condemnations of the survivors of the Shoah (catastrophe) who gave contradictory opinions of someone's conduct, of the ways in which others had been killed, and, in rare cases, who confused the faces of their torturers.[11] The fault is not theirs, but lies with a certain method of argument on the part of the lawyers defending war criminals. They demand precise statements of facts, and in this way deny that in the concrete process of remembering, facts are enmeshed within the stories of a lifetime. Facts which are part of a *story* are none the less *true*, though a lawyer might want to dispute this claim. A lawyer's case is after all merely another kind of story: a narrative structured by the language of juridical discourse.

This practice of hearing eyewitness accounts of crimes against humanity was first used in the proceedings against war criminals after the Second World War. In many countries, institutions specializing in tracking down war criminals were set up. Researchers in The Netherlands used oral testimonies from the beginning, and the same is true for many other occupied countries. By means of oral sources they documented crimes whose existence had been denied and played a major role in many prosecutions. Sources were used in a particular way. But the method, the interview, is now turning against itself. Because 'real facts' and 'real proof' are required by the judge, the focus of the work has always been on recovering the facts;[12] hence this particular use of such accounts. Other ways of interviewing can give us information of another kind, yet research on the quality of different forms of interviewing is only beginning. Also, juridical discourse dismisses the question why all this happened; why people could kill each other so brutally. It was left to philosophers like Hannah Arendt[13] to deal with the search for the origins of this distortion of human behaviour, and the question of how the mind can be attracted to atrocity. But every story is a patchwork of tales based on the suffering of a human being amazed at so much incomprehensible evil.[14]

I have often wondered if we can really understand what surviving a camp means, since the word 'survivor' has been applied to so many people.[15] If everyone uses the term for him or herself the specificity of the grief and mourning gets lost. This is true for Jews and for many others; the melting-pot equalizes and therefore denies any understanding other than a superficial denial of another's pain in a desperate effort to stress the unbearableness of one's individual fate. But no one who listens to a survivor of the massive annihilation carried out in German concentration camps will easily compare his experience with any other suffering we know of. Although recent oral history in the Soviet Union and in other totalitarian systems is certainly recording accounts that can be compared with the suffering in the Nazi concentration camps, the experience of Auschwitz and other extermination sites remains unique in its systematic industrial scale. It is in these accounts in particular that at some moment speech stops, past and present merge, and language becomes inadequate to confront the images appearing before the eyes.[16] Added to it is a complete emptiness surrounding the years of taking up a new life in a world that for so long did not want to hear, or could not hear. It is here that silence becomes a painful noise, and death is everywhere. But my work has—and this has astonished me—also been a study of the strength of the human mind.

It is not the task of oral historians to give the kind of evidence required in a court of law. Some are more inclined to study the social history of daily life, while others prefer to study the formation of narratives, and the way in which suffering is remembered and influences all other memory. In the first case, one only deals with tiny bits of daily life in a large and complex camp society divided by many relations of power, while in the second case, the focus is on memory and narratives. In both cases one is dealing with an effort to create a new kind of history that cannot be used as legal evidence since it explicitly records subjective experience. It is sometimes hard to explain to those who have been interviewing survivors in a more 'traditional' way that there is a different kind of evidence, a difference also reflected in the non-structured ways of interviewing. Oral history gives space to the narrator to reflect, to return to memories and contradict them.

My own work on the slaughtered Jewish community of Amsterdam, for instance, started with a very unconventional interview with Maurits—as I call him. He was one of the Dutch Jews of whom 80 per cent were killed during the war. I mention this, since many people are largely unaware that the death rate in The Netherlands was so high, and comparable with Poland. It meant that I had a hard time finding surviving Jews, especially Jews from poor backgrounds. It also confronted me with the task of writing history in a society where feelings of guilt are enormous, but hidden in a layer of rage against the Germans, who deported the Jews. Although there were heroic stories about resistance, which did exist, Dutch society was not able to rescue the Jews, but co-operated in a bureaucratic way. The ways in which the other Dutch stood by is most often discussed in words that ask for understanding of the need people had to survive.

There were only nostalgic works on the Jewish poor, and since nostalgia gives a static picture, the whole society seemed to have been static. It was with Maurits that I became aware of a dynamic society that was changing in many ways. He was a hawker in one of the city's markets. He attracted my attention by his speech and we became partners in an interview that lasted for months. We met regularly and I tried to persuade him to write his autobiography with me. He never succeeded in telling the same story twice. But it was precisely this that made me aware of the importance of listening and accepting contradictions. He taught me, by denying at the very next meeting what he had been saying the previous time, that he was never lying, but drawing on different parts of his memory. It was my task as a historian to explain why the stories would change, and to give them a structure. For this severely traumatized person, as for many others, the relation between life and death, day and night, was forever volatile. Memories would surface and disappear, weaving subjective patterns out of the past.

Another example of the narrative character of the life story, full of subjectivity and symbolism, came up in an interview about the diamond workers' trade union. We discussed the enormous office which was nicknamed The Fortress. My interviewee told me how beautiful the building was, and how the workers felt that this beauty had been won by the working class as a result of their union. He described the ornamental staircase, with fountains and tropical birds, and how stately trade-union officials would climb it in the company of their ladies. This interviewee lived in Israel. Anyone seeing this building might well find it very beautiful, but there could never have been fountains, and tropical birds would probably have died in the drafty halls. Anyhow, birds would not fit the aesthetic conception of the building. Since the late 1930s he has lived in the desert where water is the greatest luxury. No one there cares about birds; but in the old times, he had known poor people who had a little bird as a special pet. And both symbols of wealth and care were his way of describing his pride and love for the union.

The emphasis on evidence, in the older juridical interviews, has also created a situation in which the multiple meanings of words are no longer questioned. But if we do raise such questions, we come to astonishing results. Events and objects from the past are often traceable in layers of memory, and experienced in words stemming from the language system of that time, a language system no longer in existence. Yet, if these words are there as a defence against the emotional effect of remembering what exactly happened (very often a condition of survival), they are nevertheless transmitted in our own language system.

Let me give an example of the confusion this can create. It arose in an interview with Clara, who lived in Haifa. She emigrated to Israel after a long imprisonment in Bergen-Belsen. Before the war she had been a social worker, dealing with very poor children.[17] I travelled to see her on the extremely hot drive from Jerusalem and was supposed to meet her in the late morning. When I arrived I was tired, thirsty, and hungry. She was well aware of this and welcomed me tenderly. Before lunch she told me about her past and how she had lived as a teacher in a village outside the town. Sometimes she had had to walk for two hours in order to get home from school. If she had forgotten to bring her kosher food, there was nothing to eat, because nobody else in the village would stick to the religious laws concerning food. And as she told me, she would be 'hungry'.

After her liberation in 1945 she had written a book about her murdered pupils, which has the character of a monument. It is a very nostalgic book, in which filth has a beautiful appearance and poverty can make a child beautiful. There is no mention of hunger or thirst; the children seem only to long for love, the feeling of safety, and the light of God. I was interested in her awareness of the degree to which those children had known hunger. I knew that they had. The need for food had figured in many stories I had heard during the interviews, and hunger seemed to have been normal. In the beginning this had shocked me, since I knew of the immense charity network for the Jewish poor. And in other parts of Amsterdam people had never starved. But from the moment I began interviewing I became aware that, on the one hand, hunger could stand as a metaphor for a nostalgic longing for the happiness of childhood, while on the other, it was also used as a metaphor for the war experience. It was part of a standardized narrative. Its mere existence in the stories had everything to do with the war, I thought. But how had she reacted to hunger, since she was not often hungry before the war? Clara's answer was:

Oh there was hunger, but not the way I have known it since.

Can you remember——

I stood in Bergen-Belsen at the dustbins. If there were people who could not swallow what they gave us, and wanted to throw away that nasty food, I would ask them to give it to me——

But even if the hunger was not as great as in Bergen-Belsen——

I was in Bergen-Belsen.

No, I mean in Amsterdam, so earlier—did you see much hunger even if it was of a different kind?

Yeah, but we did not know then what hunger was.

I knew I was acting against all the sensitive rules regarding interviews, but I wanted to know. Although she had depicted with love the poverty of the ghetto, she could not imagine any other form of hunger than what she had experienced in the camp. When we walked away to get lunch she worried about me. 'Aren't you hungry?' she asked. Later that day she wanted me to stay in order to prevent me from becoming hungry.

It was very clear that Clara could not talk about the extent to which the children she had worked with were hungry, nor could she give any other information about their desperate lives, other than by using the official language available at that time for charity workers, saying that they were good children and only needed a little help. The children were all dead; she had tried to prevent the deaths of other children in Bergen-Belsen, and she had failed. That was how she felt. In her story of sorrow, the children turn up only as miserable creatures who needed protection. A first impression is that this story was told only in the language which certainly, as I gathered from written and oral descriptions, had not been hers at the time. Her success in creating a house for those children had been the result of her lack of distance, her immediate solidarity, and her manner of speaking, so different from those of all the other professionals.

In a court, Clara's eyewitness account would have been attacked immediately. Her facts were as unclear as her love for the children was clear. If she used terms such as 'forlorn streets', ten witnesses might have been able to describe how nice the streets had been to play in; only inside the house could this have been the case. She described the water in the canals as dark, which might be a reflection of her mourning, since in this part of the town, for centuries, the water has been very clean, due to the system of locks that connect the canals with the sea.

Just as this interview would not stand up to a trial, being so overshadowed by the feeling of loss, it would also fail to stand the scrutiny of traditional historical research. There was too much of a mixture between present and past images in the interview. It was a hopeless effort to transmit some notion of a completely devastated culture to a much younger person. How could she transmit the noise of the ghetto to me, who had never lived in a ghetto and who was raised in a country where the Jewish population was diminished to a greater extent than in all the other countries of western Europe? How could she transmit the safety she had felt before the war, rooted in a tradition of tolerance, when we were now so brutally aware of the fact that nowhere else in Europe had the bureaucracy adapted itself so easily to the new regime and collaborated so smoothly with the Nazi authorities?

Clara was one of the many people interviewed for my book on the Jewish proletariat of Amsterdam before the war. The book is based on a series of ninety interviews; some were informative, some were real-life stories.[18] I tried to come to terms with a nostalgic image of the past, penetrating everything that had happened before the great evil. It was, I felt, my task to confront the audience with a more challenging history that would really question the problems veiled by nostalgia.

In daily life we have ways of dealing with our memories: there are members of the family to correct us; there are streets and signs that make us reconsider the past; and there are of course newspaper articles, television, and official stories. But this was not the case in my research. The old ghetto has been physically obliterated: moreover, other neighbourhoods where Jews used to live have changed completely. A whole culture had disappeared. A view of the past is structured through the personal loss of family members and loved ones, and the destruction of the cultural co-ordinates of personal lives. The world that had once been there was a world of another time, with other values, to be longed for.

Of the 80 per cent of Jews who were killed, the majority were proletarian. They were victimized before the German occupiers attacked the richer classes, and had fewer means of escape. With the trains to the east the Jewish proletariat ceased to exist. I decided to interview those who were still there, as an academic project, but the work was not undertaken out of intellectual interest alone. I had often wondered about the extremely nostalgic romanticism of the literature on the old Jewish neighbourhood. It seemed as if there were no other way to deal with the past. And I doubted those stories, knowing of others—bitter ones from sources I had seen in the archives. I had also heard stories of desperation from people who had been brought up in the slums, where they had slept on paper, under rags, and lived in one-room apartments with big families. Who were they? Were they as socialist as many tales would have us believe? Who were those pedlars and hucksters, and would it be possible to remember them differently in a life story? Would it, with this method, be possible to recreate a more realistic picture, and one that was less static than the images created from nostalgia? What were their struggles, hopes, anxieties, and doubts?

Most of the Jews who lived in Amsterdam before 1940 were poor. They formed about 10 per cent of the city's population. Although they had enjoyed formal equality since 1796, in the twentieth century they were still excluded from a large number of occupations. Anti-Semitism in The Netherlands was less virulent than in many other west European countries, but social and economic networks frequently excluded Jews or admitted them only sporadically. The working class and paupers mainly worked in the diamond industry, textiles, tobacco, and trade.[19]

Before the turn of the century, most of the Jewish poor of Amsterdam lived in clearly defined neighbourhoods, where living conditions were very poor, and where people lived according to their old traditional ways. I researched the efforts to 'civilize' people living in this culture into 'normal' ways of life, in the context of the particular character of the economy and the influence of socialism. More than any other group, the Social Democrats involved in the making of the city's politics carried out the policy of civilization; new houses were built, resulting in the construction of a new working-class district in the east of the town. Jewish working-class families who moved into the new quarter or other new housing had to adapt to a culture that was based on a socialist conception of the modern nuclear working-class family. This was not in accordance with many Jewish traditions such as strong family networks, living within an extended family, distinct cultural expressions in street life, and a different way of managing the household. In the years just after the first World War many Jews had tried to adapt to these new ways, since the programme of modern life attracted many socialist Jews. But in 1929 the Jewish economy was harder hit than other sectors, and pauperization became general. People clung to the old ways of living, and no longer believed in the new times of socialist promise. But since they had believed in it, the result was ambivalence about Jewish identity and values. This added another layer to their complicated relation with Dutch society, which in turn was overlaid by their ambivalent feelings about returning to an inhospitable society after the war.

Maurits, mentioned above, was an example of the ensuing ambiguity of Jewish identity. He was deeply ambivalent about the degree to which he wanted to integrate into Dutch, non-Jewish society. He was proud of his socialism, and had spoken about humanity instead of Jews. Differences between people and nations belonged to outmoded ideologies. But he had also enjoyed Jewish life: candles on Friday night—a world he wanted to talk about. His story was about his present ambivalence, a confrontation between his former hopes for a better world and the knowledge that a Jew is always a Jew in the eyes of others. He revealed to me the different stages of his life and his different feelings about them. But alas, his memory was not only a memory, but also the traumatized narrative of a survivor. Those traumas were more apparent than his doubts about his place as a Jew in Dutch society. Oral history is a major resource in researching how present doubts about identity interfere with the past, since it offers a perspective for research into long-term cultural patterns.

Death and murder, always present in the stories, were emotionally exhausting for me. It was clear that one way of facing their cruelty was through a professional attitude, stressing again and again to myself that I was searching for historical knowledge. The very same attitude was needed to face the other problem that turned up repeatedly: that of Jewish identity. No one has a clear view of the place of Jews in the history of a country from which so many were deported. The integrated Jewish proletariat had seen how their comrades from the unions and their neighbours had looked away. The uncertainty stemming from these experiences was not something I could share. But of course they wanted to know how I felt as a Jew in such a country, and we both knew how opinions could change over time.

The relationship with Maurits confronted me with both problems for some time. It was hard to avoid playing the role of a social worker, since he needed my attention so much and was so delighted to have found in me a silent listener. There is a kind of understanding a historian can offer that nobody else can. We know where the streets were, we know about the policeman on the corner and the sweet smells from the bakery. We can answer questions about the name of the street, the shop around the corner, or the name of a well-known baker. With historical knowledge one is able to listen in a specific way and sometimes discuss confusions of memory. This was the case with Maurits; my detailed picture of his environment and culture made me aware when Maurits was contradicting himself, and I could often understand why he did so. It was particularly interesting to see which story belonged to what kind of memory, and accept that there was no final reality, not an ultimate truth to be reached. But this does not imply a dismissal of the interviews as mere fantasies that give no accurate historical information.

The specific way in which a historian listens came up vividly during an interview with a well-known psychoanalyst who had pursued his career after the war. He kept repeating the story of his childhood, a lovely story that situated him in the poorest part of the working class. It was immediately clear to me that he could not come from there. This was apparent as a result of my knowledge of the street where he lived, his father's job, the school he went to, and so on. He told me about the horrid smell coming from a house he had to pass when he walked to school. After the interview I wondered what kind of smell it had been and looked at a map. It was clear that he had not passed any butcher's shop, factory, or warehouse, but only ordinary houses. And I wondered how much this memory was part of the other errors. I went back to show him that the factory he had described did not exist, and I told him—assuming that with his training he could take it—that there was no explanation for this smell. In normal interviewing one would never do that, but he was so exceptionally well trained in memory that I assumed I could raise the question in an intellectual debate. At that moment his childhood collapsed, and also his psychoanalytic reconstruction of it. He had told this story over and over in his therapy and he had believed in the story. He became aware while we spoke that he had mythologized his childhood recollections in order to emphasize his upward social mobility. In analysis he had never encountered anyone who was interested in whether or not the smell was real, since psychoanalysis is not concerned with that kind of realism.

As soon as we begin to see to what extent the stories are reifications of the past in the present, we also start to have a glimpse of the past itself. I interviewed Mrs Fré, who had worked in an orphanage. In the written sources this orphanage has been extensively described. Observers tell us about the modern furniture, the strict dietary laws, the religious training, and so on. I had been interviewing Mrs Fré for a long time before I realized how much these pictures were part of a romanticized view seen through the eyes of the trustees. I was glad to finally have found someone who had worked with the children. Mrs Fré had looked after the smaller children and had been very idealistic about working with them. Her biggest complaint was about the shaving of the children. As she put it: 'If you had that sudden rupture that your father died, and the great poverty came, then it was decided that the children would be sent to the orphanage, but they went to a hostile world. Away from home, to which they sometimes felt very bound, and then—because sometimes a child had lice, he was shaven—the paraffin cap stayed on for three weeks.' She wanted to do things differently, in a more 'modern' way, as she put it. During the interview I did not get much information that could be considered new. Listening again to the interview and reading the transcript repeatedly, I became aware that she had not wanted to tell me about the orphanage. She had only one message, which that she loved the children and yet could not prevent them being killed. This was the only message she had for me, and only after I had made clear that I understood her was an interview at all possible.

This could have been the end of my interviewing, or rather failed interviewing, about the orphanage, had I not met one of the orphans. He wanted to tell me how horrible the institution had been, since he felt that all the written information was too romantic. His story made me understand Mrs Fré better: 'It was a terrible moment when I was brought from . . . to the orphanage, but all the other unpleasant things happened at once. If you came into the orphanage in those years they assumed that your head was not clean. The first fourteen days you got a paraffin cap.'

His whole story was about old-fashioned severity, in an institution where children were washed with cold water. Punishments were harsh and the atmosphere was cold. But at the same time he told me that the trustees had seen to it that he could go into higher education, something very uncommon for a child from an orphanage.

His story stands in marked contrast to the story of Simon, who never had any opportunities, being a child whose parents could not take care of him. He was brought to a more general institution and was made to eat pork. The feeling of being deserted has never left him. The Jewish community did not take care of him and he had to struggle alone. His account gave a wealth of information on the institution and the aloofness of the Jewish community, the latter being a story never told in the official literature. The difference between the two stories is included in the other message that Mrs Fré had wanted to tell me. She had known that the system was old-fashioned and she had wanted to change it. She had cared for the children, since in most other institutions no one seemed to care. She was angry that the negative side of the strictly religious and hierarchic but very protective life was neglected in the memoirs she had read. At the same time she knew that some children obtained exceptional chances, and she had worked for the improvement of the institution. But she had never succeeded. With this in mind the interview can be read as a résumé of modern pedagogic criticism of a system. It also provides a tremendous amount of information about the system itself. At the same time the interview is a document about her grief.

Another typical example of the kind of information available to the oral historian about such a history of mentalities can be found in an interview with the 92-year-old Sara—as I shall call her. She was not born in Amsterdam, but arrived there in 1910. She did not like the outskirts of the ghetto she had moved into. She had been amused by the street life, but also despised it. Married to a diamond cutter, she later found a home further away from the ghetto, and she described her feelings in the following way: 'When I arrived it was a revelation, all those street people. We did not have that in Rotterdam. After a time you get used to typical Amsterdam things. It was so different from back home.' She had moved house and I asked if the new house had been better. Her answer was negative, but the others living there were better: '. . . a different neighbourhood. It was also Jewish, but not that very Jewish.' In these and the following words Jewishness stands for the ghetto, which in her story is plebeian, dirty, and backward. I tried to clarify the distinction and asked about the Jewish character of the first house: 'Yes it was very Jewish; those women were cleaning vegetables in the street. They sat near the door, and I was not used to that.' I was curious to know if Jewishness had other meanings than poverty and ghetto life. Very soon it turned our that there was also a positive side to it, which had to do with keeping up tradition in a modernizing society. All her friends were Jews, but not Jewish, which was also a question of class.

The positive side of being Jewish is traceable in the reaction to the suffering of Jews and their estrangement from the non-Jewish world. Positive feelings about Dutch society were replaced by feelings of betrayal, especially after Jews came back from the camps. I quote Sara again:

> I came back in a cotton dress and a pair of soldier's shoes. They never asked me: 'Do you have a dress, or a house,' very few . . . Nothing, nothing, nothing. While I came back very late. I only came to Amsterdam in October, because I was so ill . . . Then you could not get a house, because everything was inhabited. You could get lodgings formerly inhabited by the NSB [Dutch fascists]. But they did not do much for us.

This bitterness about their reception in The Netherlands is very common among Jews who came back. They were not welcome, and modern research suggests an upsurge of anti-Semitism in the first months after the war. People wanted their houses back, and their belongings, but they found a guilty hypocritical society, in which even tablecloths that had been 'in hiding' were often kept by those who had offered to take care of them. Insurance companies harassed the survivors about overdue payments. The building of a new life through remarriage was not an option because it was seldom possible to prove that a former wife or husband had been killed. This actual estrangement is mentioned first in the interviews; it is more visible than the mental separation that Jews felt from the rest of society. But the stories give a glimpse of a more violent trauma: the experience of hostility from the very society into which they had previously so successfully integrated.

The lonely return of the few survivors—a mere 5 per cent of a once strong and thriving community—to a society that did not welcome them and did not give them back their houses was often experienced as worse than any humiliation suffered during the war. A romantic image of The Netherlands had kept them alive for many years. But the survivors came back to a society whose people wanted to forget and were laying the foundations of a different collective memory. Survivors could not forget, and did not want to: so the behaviour of the Dutch shattered their memory of the pre-war period. They had once felt accepted, but they did not feel that way any longer. Objective isolation lasted until the 1970s when their severe psychic problems became generally accepted, but for many of them it was too late to break down the wall of self-defence. They had also had to fight for financial support, and only very recently have Jewish journalists and historians raised questions about who the policemen were who took people from their houses, and who drove the trams to the railway station. Dutch society has never been able to face the moral question of this part of its wartime history. It was not discussed; but the silence was deafening. Thus even pre-war memory was influenced not only by the war but also by post-war attitudes.

Louis had been full of the promise of integration. A diamond cutter who had not known very much about Judaism, he now regretted his ignorance. In the interview he stressed that it would have been better if things had been different, if he had known 'that there really were Jews', and not just human beings. In the train to Auschwitz he became aware that he was being deported for a reason he did not understand, since he did not know what it was to be a Jew (I remember protesting that even if he had known, this was no reason for his deportation!). He had thought that he was no different from *all* other people. Very soon it became clear that he could not have felt very different, because he seldom spoke to non-Jews. His school had been overwhelmingly Jewish, and so was the place where he worked: the diamond industry. Even his free time was spent with Jews; and, he talked to them in the barber's shop.

He said again and again that he did not live like a Jew. His shame was apparent. According to him he started to become a Jew in the train to Poland. Apparently his idea of being a Jew had become religious, although he was not religious when I interviewed him. But in his eyes he was a Jew now, above all because he had been in a camp. At the same time, his picture of a pre-war life spent exclusively with Jews was a major element in his mental story of assimilation. Jewish identity was consciously changed and fragmented by the deportation and by the changing relations with non-Jewish Dutch society.

Interviews are complicated by more than the effects of the war, and it is not only nostalgia about the pre-war period that creates the contrast with what came afterwards. The massive deportations were a tragedy so devastating that a historian can easily overlook the way in which survivors may also be reacting to things other than the war. Their lives might in any case have been fragmented. One example of how patterns can persist and confuse the researcher, even if they have nothing to do with fragmentation, is given in Simon's story.

Simon was clever enough to escape from the Germans. He did not even go underground. At the end of the winter of 1944, during the period of starvation, he went to look for food north of Amsterdam. He managed to escape the controlling patrols when he came back with the food.

Let us take a look at Simon's success: as a small child he was sent to an institution for homeless children and adults. He managed to find his way there; he seems to have been very good at it. Later, in the 1930s he was very clever in avoiding control by the social security. His escapes formed a pattern that enabled him to survive. Throughout his whole life he has had to use this ability. What he did during the war was indeed very clever. Does that create a pattern for the period before the war, or did the pattern make him interpret his life before the war in this way? Or did this pattern exist before, and was it precisely this skill that helped him? With Simon these questions are further complicated by the fact that he described his life as if he were currently doing well. But it was obvious that he was not.

In the process of remembering, several levels of language are deployed. The meaning of words is often highly personal, but still they are not mere inventions in the speech acts of individuals; they also reveal the person's efforts to come to terms with the contradictions in his or her history. My narrators took great pains to find ways to express what was essential. They defended their individual accounts of their lives and vocabularies against the stories they assumed I knew. And at the same time they anticipated questions they expected me to ask. The stories were shaped according to a collective memory that was supposed to be part of what they thought I expected. I became in a way the revenging authority of the super-ego, to whom they tailored their narratives, and to whom they expressed their understanding of what had been repressed in the collective consciousness and official history.

Many details about Jewish culture and life have turned up precisely in the form of anecdotes. It turned out to be possible to use small details to determine the social position of the family the interviewee stemmed from: for example, the fact that they would only take a shower in school, or would visit a public bathhouse. The ritual bath for women is an indication of their religiosity, as is attendance at religious schools. These are the small details that give us a wealth of social and historical information which merges into a history of mentalities.

The most essential element in all these stories is that being Jewish has become for them a confrontation with death, but this was not always the case. In this sense, oral history with survivors is different from any other kind of interview. Death is always present, and it has become an aspect of the memory of the time before the war. There had been no farewell to this past, symbolized in the sudden separation from parents: the normal process of grief has become sorrow about anonymous graves. Any balance has become impossible and has been replaced by the language of nostalgia, as shown in the glorification of a reified traditional Jewish family life which does not correspond to the enthusiasm with which they had greeted modern, assimilated life. It was only after many years of interviewing—and running up against this contradiction—that I came to realize that the word 'Jewish' stands for something that has never been relinquished: it symbolizes a clash between two cultures; one of them had become the inevitable winner.

Some people became more Dutch and less Jewish as a result of the war; others did not. For others, things remain ambiguous. I think oral history and the use of life stories can bring these confusing contradictions to life again. But narratives of the interviewed survivors are also constructions of the present. They are structured and biased by a whole series of desperate experiences throughout a lifetime. This becomes a problem in itself, as soon as we deal with the war and the Holocaust. How do we relate our results to the field of research dealing with the ways in which different non-Jewish individuals and groups came to terms with that period, and have created forms of collective memory which might conflict with or even exclude the memories of others who witnessed the very same events? There clearly are worlds of experience and collective memories that cannot be shared by others. It is precisely such collective group memories that can become part of a national memory or be excluded from it.

In Dutch collective national memory, most Dutch people were anti-German. The writers of the history of the Second World War emphasized this democratic character of the Dutch population until about ten years ago. Now, a new generation of historians has started to reconsider this idea. It might seem natural that the resulting research would have brought about a critical re-examination of Dutch attitudes and behaviour. But this has not been the case. Recent historical writing argues that the bureaucratic attitudes of the Dutch population were principally due to the simple wish to survive. People had to eat; and many did not believe in the first years of the occupation that the Germans could be defeated.

This is a moral issue, very similar to the historical evaluation of the behaviour of a small *apparatchik* in former communist countries, or the behaviour of the German working class. But moral positions are sometimes presented in terms of a more scientific approach. There have been too many accusations, it is suggested, and accusations are not the task of a professional historian. In my research, I also found many cases where people tried to explain that the Dutch national myth in which everyone was on the right side was a distortion. They knew better, since they had seen many people turning away when danger was really present. That had made them think about their position as Jews in The Netherlands, and had made their hopes of integration seem ridiculous. In this sense Jewish memory is a useful tool for research on the Dutch mentality of the war years.

It seems that the historical debate has also become a debate about collective memory. The new history excludes Jewish memory, or makes it pathetic. It is, in my view, primarily a way in which a guilty society can give itself a positive identity. It is known that the war period was problematic, and that not all the Dutch were on the right side. Jewish memory has always known this, but has led to different conclusions about the place of Jews in Dutch society. Life stories are fragmented by hopes of integration and acceptance, and their disillusionment. But the confusion is made greater by a fundamental uncertainty about Dutch society. One major task of oral historians might be to confront contemporary Dutch historians with the language of the witnesses, and show its value to the academic world. Oral histories can demythologize collective and official readings of the past, even if these readings are constantly reframed. Survivors have been forced to create a juridical language; but they could create another. Their voices would be in strong opposition to much modern academic writing of history; for in the final analysis the writings of historians are characterized as much by what they exclude as by what they explicitly say. Recent Dutch historical writing, while claiming to be more modern and more academic, has left no place for history's victims. The debate in that sense is a political one: does the suffering of Jews still belong to Dutch history, or has it become instead the history of others?

Notes

1. S. Leydesdorff, 'The screen of nostalgia: Oral history and the ordeal of working class Jews in Amsterdam', *International Journal of Oral History*, 7/2 (1986), 109–16.
2. An excellent example of the way in which especially the trial of Klaus Barbie gave rise to a genre already traceable in other Holocaust literature and using oral

testimonies is, T. Morgan, *An Uncertain Hour: The French, the Germans, the Klaus Barbie Trial and the City of Lyons, 1940–1945*, London, 1990, where the major focus on the trial leads to a history of events, in which the boundaries of juridical discourse are not overcome.

3. G. van den Berghe, *Met de dood voor ogen: Begrip en onbegrip tussen overlevenden van Nazi-kampen en buitenstaanders*, Berchem, 1987, 363.
4. A. H. Rosenfield, *A Double Dying: Reflections on Holocaust Literature*, Bloomington, (Ind.), 1980.
5. P. Levi, *The Rewakening*, New York, 1965; and id., *Survival in Auschwitz*, New York, 1969.
6. V. Fränkl, *Man's Search for Meaning: An Introduction in Logotherapy*, New York, 1964.
7. P. Célan, *Selected Poems*, London, 1971; and id., *Speech-grille and Selected Poems*, New York, 1971.
8. The major historian of this kind of work is R. Hilberg, *The Destruction of European Jews*, New York, 1961. Problems of the very expanded field are discussed in R. Marrus, *The Holocaust in History*, London, 1988.
9. B. Bettelheim, *Surviving and Other Essays*, New York, 1980. See also 'Il prezzo della Memoria', in A. Bravo and D. Jalla (eds.), *La vita offesa: Storia e memoria dei lager nazisti nei raconti di duecento soprawissuti*, Milan, 1985; F. Crejeja and B. Mantelli (eds.), *La deportazione nei campi stermino nazisti: Studi e testimonianze*, Milan, 1986.
10. The chief example is E. Wiesel, *Night*, New York, 1960.
11. The trial of John Demanjuk is the most frightening example of this. See T. Teicholz, *The Trial of Ivan the Terrible: the State of Israel versus John Demanjuk*, New York, 1990.
12. On the problem of juridical use of history see A. Portelli, 'Oral history, the law and the making of history: the "April 7" murder trial', *History Workshop Journal*, 20 (1985), 5–36. Of course, one cannot compare trials of Italian terrorists with those of war criminals. But Portelli too comes up with the problem of historical memory and knowledge used as a tool in the accusation. His conclusion is that memory cannot be used in court as it has been in those trials. I came to the opposite conclusion in the specific case of war trials. This could be an interesting field for research.
13. H. Arendt, *Eichmann in Jerusalem*, New York, 1970.
14. Unquestioned so-called eyewitness accounts have been used for many purposes, such as the anti-fascist rhetoric of the former German Democratic Republic, and for the denial of France's Vichy past: H. Rousso, *Le Syndrome de Vichy*, Paris, 1987.
15. The most disgraceful example of this is J. Bacque, *Other Losses*, London, 1990.
16. I was very moved by the effort I found in F. Ringelheim, *Les Juifs entre la mémoire et l'oubli*, Bruxelles, 1987. See also N. Lapierre, *Le Silence de la mémoire: A la recherche des Juifs de Plock*, Paris, 1989.
17. C. Asscher Pinkhof, *Sterrenkinderen*, The Hague, 1946 (trans., *Star Children*, Detroit, 1986).
18. I started with an attempt to imitate M. Zborowski and E. Herzog, *Life is with People*, New York, 1952, which in its love for daily life has been a source of inspiration. It is also an attempt to use testimonies in the description of a lost culture.
19. See also S. Leydesdorff, 'In search of the picture, Jewish Proletarians between the two world wars', in J. Michman (ed.), *Dutch Jewish History 1*, Jerusalem, 1984, 315–5; and, 'The policy of the Amsterdam municipality towards the Jewish proletariat in the 1930s', in Michman (ed.), *Dutch Jewish History 2*, Jerusalem, 1989, 235–47.

10

Don't Forget

Fragments of a Negative Tradition

RENATE SIEBERT

> She was seated, her hands on her lap, moving them once in a while by intertwining them as if she was playing. Her face had the wide-eyed astonished look of someone who has just woken up and cannot yet recognize what she sees. When someone talked to her, she put on a naïve and sweet smile, full of serenity and almost of gratitude. But no one could expect any answers from her. As a matter of fact, she seemed hardly capable of hearing the voices, and could not understand either a language or, perhaps, a word. Sometimes, with a dreamlike mumbling, she would repeat to herself some vague syllables that seemed to come from some dreamlike or forgotten language. With blind and deaf people communication is possible. But with her, who was neither deaf nor speechless, no further communication was possible.
>
> (Elsa Morante, *La Storia*)

What is our relationship with the past; with our recent Fascist and National Socialist past, which is a heavy burden on our consciousness, and which shadows our historical and social memory? Who are we, as individuals, in relation to this past? Do we have choices in the face of what the past forces upon us? What strategies do we have to face the past?; to remember?; to forget? Are there others . . . ?

Iduzza, a quiet and at the same time a stubborn woman, scarcely manages to get through her daily life, as she is gradually overwhelmed by the unceasing barbarism: helpless, herself a Jew, she witnesses the deportation of the Jews from the ghetto of Rome. Terror, a monster with countless heads, invades her home, her entire life. Finally it steals her last hope: it kills her son Useppe. Her loss of memory, madness,

Translator: Alessandra Lorini.

marks the collapse of her consciousness in the face of the immeasurable and the unspeakable. By using the words of her condition, by stretching and compressing her language, the main character of *La Storia* communicates her horror to us. We are made participants both in her inability to withstand anything more, and also in her denial. In contrast to suppressing or forgetting, her loss of memory reflects a refusal. It is a desperate but firm refusal to accept shared values, and join a foreign world.

A very different relationship with the past underlies the process of suppression. It is a process of knowing nothing, looking the other way, refusing to recognize, and, most important, of expelling all that is problematic for one's own mental universe. This denial and pushing-away of whatever is painful and uncomfortable can be individual or collective. Suppression is built on the illusion of eliminating what is disturbing. By naïvely closing one's eyes, one hopes it will prove a mere nightmare by the time of awakening: obsolete, trivial, and inoffensive. Often, this rejection of the responsibility of remembering is linked with a more general tendency of self-absolution. On the other hand, to keep memory alive, not to forget—the Jewish '*zahkor*'—reveals the existence of an ethical relationship with memory, and of a will and a need to elaborate the process of mourning. It is both an individual and collective sharing of the unhappy destiny of the victims of totalitarianism. 'We are all guilty', Sartre said; 'We are all Jewish', we used to cry in student demonstrations in Germany in 1968. Not to forget is not meant to be a ritualization or fossilization of the past, but a way of confronting ourselves, day after day, with the desperate promise 'Never again'. The memory of the past, rather than a shadow over the present, thus becomes a part of it, adding vital depth, transcending the element of mourning. In this light tradition takes on a special connotation. It is no longer an indisputable authority or a set of rules, norms, hierarchies, and ultimate beliefs, but a mediation between the fragments of a collective past and the consciousness of individuals. It is a sort of negative tradition if one could possibly define it. What is, in fact, tradition? Is it that relationship with the generations that have come before me and the institutions that they have left? It is the past that comes near me, touches me, absorbs me, and surrounds me. But there are ancient traditions and more recent ones. There are dead traditions and those that are still alive; fossilizations and caricatures of traditions. Furthermore, there are good and generous traditions, and those that are bad, evil, stenching, or deathly. What is the relationship between institutions and traditions? Do traditions select, save the good, obscure the disturbing and deathly? Do they lead us or do they deceive us? What is the authority of traditions and how do they affect the individual? Is it possible to skip recent history, the history in which our fathers and mothers were protagonists—because it is rotten—and connect to previous traditions? I do not think so. Can 'not forgetting' become an uncomfortable, yet living, radical, and accepting tradition? I believe so.

I feel the need to talk about myself; to describe the hard and painful path I have followed, which was strongly marked by the fearful and almost unspeakable memory of German National Socialism. I should say that what I want to talk about has had a pervasive and violent impact on me. I have felt disturbed since November 1989, when a particular event was reported and commented on by the Italian press just like an ordinary daily happening: the showing of documentary film, produced by the BBC, entitled *Fascist Legacy*. It offered a historical reconstruction of the war crimes the Italians had perpetrated in Yugoslavia and Ethiopia. These crimes have remained unpunished, thanks to the will of both the British and the Americans. But they have been forgotten by the Italians as well. I cannot deny that I was made very angry by the reception given to this documentary by the Italian media and public opinion (of which I will say something later). I am disturbed by my anger and I would like to make it more 'objective' through my writing.

I was born in West Germany in 1942. I belong to that generation of people who during their adolescence suffered the trauma of seeing the collapse of all points of reference and credible authority, both on a collective and family level. Like the child of the film *Germany, Pale Mother*, I grew up among people who had been emotionally destroyed by the war, by bombings and by the loss of the illusion of grandeur of the regime which they had more or less willingly supported. Even this, however, was not known, as it was concealed under the thick layer of anguish-generated silence that had swallowed our recent past. Today, if I reflect on the lack of elaboration of the connections between my own family's story on the one hand and the political and social history of National Socialism on the other, it feels like being sucked into a vacuum. Once again, Helge Sanders's film comes into my mind: depicting a destructive fury that pervades emotional relationships, and a silence that, by exploding on an emotional level, like a torrential river carries everything away. My parents were like this, and this silence has reached me: a disturbing silence, the silent anguish of war and its destructions, the total absence of hope, and a dull and stubborn self-destructiveness; brutality and self-defence; extreme fragility.

As I remember, anti-Semitism was the turning-point that, like a blade, cut through the apparent immobility and discontinuity of inter-generational relationships, and in the mid-1950s led to a rapid change in school programmes. Suddenly, after the embarrassing silence, came 'information', data, and figures: how many millions of Jews, Russians, and gipsies had we exterminated? Twenty? Thirty? No, only eighteen? We looked at the evidence, at those documentaries that filmed the arrival of the Allies in the concentration camps. These were unforgettable, unaccountable images of a surrealist nightmare. Even today, whenever I see the transit of a freight-train in a railway station, I experience a deep anguish. This scene always and only evokes the images of deported Jews and of their arrival at Auschwitz.

That year marked a turning-point in my school-days. I had grown up among people who were emotionally destroyed, but I had not known it. From then onwards, I would put all my efforts into breaking my relationship with them. In other words, I believed that we had to pass through there, through Auschwitz, in order to have any real relationship. I wanted to know everything and I wanted to know it from them. But they did not answer me. Those were the adolescent years, which are unavoidably conflictual. But ours was a life-and-death struggle. And it meant the death of most relationships with the adult world. There are people in my family with whom I have not exchanged a word since then. I remember participating in dramatic discussions with teachers who would leave the room in tears, slamming the door. We had no mercy. We had been deeply betrayed.

I began to spy on adults. I was trying to catch them out in self-contradiction, discover them. We needed to provoke them. I read a lot as I was obsessed with knowing what had happened. I remember that in Karl Kraus' *The Third Night of Walpurga* I found an episode referring to a public persecution of Jews that had occurred in 1933 in my own town, and, by chance, in the very same street where my mother had lived as a girl. I read the excerpt to her, and watched her. Up till then she had maintained that she 'had never known anything'. 'Oh, yes,' she now answered, with no hesitation, 'I remember it. But you know, they were Jews.' It was almost unforgivable.

I felt guilty. *We* were guilty. I felt ashamed for them. I dreamt about war, destruction, and corpses. When I was abroad I tried to hide my origins. I would either try to speak other languages or keep silent. I was afraid. I was afraid of being 'discovered' as a German and, accordingly, held guilty. But I was also afraid of being arrested by the German police. I had an obsessive fear. An old Fascist saying kept sounding in my ears: Never foul your own nest. My turmoil found an outlet in anti-social attitudes such as small thefts, and even some car thefts, just for a ride . . . So my obsession began to make sense. I had good reason to fear the police. I was hiding from them even before I got involved in politics.

Those were the years of the cold war. I lived near the border with the GDR: that unknown continent, the other Germany, that poor sister, that ugly duckling, like a forbidden territory, suppressed from our consciousness, demonized by propaganda. It was almost dreamlike in its unreality. It was a non-place, a mysterious entity, which was neither comprehensible within its own (non-) tradition, nor definable as a foreign country. In my solitary search for origins, the GDR was the first surrogate for my missing country. I listened to its radio stations which openly spoke of anti-fascism! During my last high-school years I followed semi-clandestine courses given by a few rare groups of communists—in those years in the German Federal Republic the Communist Party was illegal. I began to react to my nightmares. I should be and wanted to be different. First of all, I needed to be strong, and to resist. I read whatever was available on tortures. The Algerian liberation war was going on, and I would read to challenge myself: Would I be able to resist and not betray? Would I be able to carry out my duty right up to the ultimate consequence? A commitment that could not jeopardize one's life was not a commitment. We had already seen this. I wanted to remember everything and never forget. I would have died rather than betray.

When I went to college I was finally able to join in socially. It was a wonderful discovery. Many of us felt like orphans. My college years were those that preceded and prepared for the great explosion of 1968. We worked together to understand what had happened; we fought battles to eliminate the residues of National Socialism in our society; and we picked on racism as the key element. Solidarity with liberation struggles in third-world countries created new connections of solidarity among ourselves too. Our unmasking of hypocrisies about the past and our struggle against racism in the present meant the same thing. We helped deserters; first the French from the Algerian War, and later the Americans from the Vietnam War. We denounced discrimination against Italian and Greek immigrants. We tried to leave our sad family histories behind us, for we had found new spiritual and exemplary fathers in our teachers: from Adorno to Horkheimer and Marcuse.

These are crucial elements, I believe, for understanding the German 1968: it was in great part an elaboration of the memory of fascist totalitarianism. 1968 gave our orphan generation a perspective on the present which also enabled us to look at the past with less disoriented and desperate eyes. We could plan for the future, we could hope. Nevertheless, when I think about individual comrades of that time, I also recall the existence of a strong personal predisposition to despair. Drugs were not available in those days, and we drank a lot. Some committed suicide. Others made seriously self-destructive choices, such as opting for terrorism. Not all of us were able to trust and recreate a credible world or establish emotionally lasting relationships. Another film, *The German Sisters*, by Margarete von Trotta, offers a convincing analysis of how much the problem of terrorism tormented my generation. Although I have made different choices, I can say that I shared many things with those who died in prison at Stammheim. The impossible past bars the road to the future.

If a stubborn and obsessive insistence on the past has characterized many years of my life, there has also been, in a rather peculiar way, an escapist tendency: the metaphor of the journey and the nostalgia of never returning. Ever since I can remember I have always thought and dreamt about leaving. When I was a girl I tried to persuade my mother without succeeding, 'Come on, let's go, let's never come back.' Then I silently built my own project. During my college years I would disappear every now and then; I would give my furniture and other belongings away and vanish. I wandered around the world, always heading south—I am afraid of cold as my childhood memories are memories of cold and ice. I would return after some weeks or sometimes months. For my comrades it was as if I had never left. We were a permissive 'family' challenging all sorts of authorities. As far as those few objects I inherited from my real family—jewels included—are concerned, this irregular lifestyle has had a devastating effect on them. Miraculously, I still have a small rococo porcelain statue, the only witness of a world that is definitively gone. I care very much about it now, and I am very careful not to lose it. At the same time I do not regret all I destroyed or lost. Evidently, this was necessary to come to terms with that awful past.

During the last twenty years I have lived and worked in Italy. I have become an Italian citizen. Some of those nightmares were so strong that I even needed to give up my German nationality, the official one that is stamped on my passport, to cope with my anguish. I needed to go to the very end. I believe I did the right thing. Now I feel freer. Sometimes I speak German without feeling shame; I can go to Germany without having nightmares, and I have forgiven my parents. I could finally bury them.

How did things go for the Italians of my generation? Which memories do they have, which collective or individual traumas? I cannot tell or, better, I have too little on which to base an answer. I can only hazard some hypotheses. I can say that, to my regret, I spied on them too for many years. I was really struck by the tranquillity with which the past and the memory of Fascism could be evoked. As a historical, political, and social question, Fascism is rarely seen as a personal problem. It is as if fascism does not concern personal identity and subjectivity. This is what I have observed through living in Italy.

One possible explanation might be that in Italy the relationship to the memory of totalitarianism has shifted onto the level of an almost institutional conflict within politics. Two elements have certainly contributed to making this 'distanced' elaboration possible (I mean distanced in terms of psycho-existential involvement). On the one hand, the blame for whatever was disturbing is placed on the Germans, who were not only evil but avowedly so. When compared to the atrocities of the Third Reich, Mussolini's regime could appear like the harmless staging of an operetta. This type of ideology is rather common. On the other hand, the partisan resistance, an undisputable but also mythical reality, offered a positive object of identification: it provided 'good' fathers, especially for militants on the left. There was never a real generational fracture—at least this is how it appears to me after living in Italy for many years—because an empty space in the transmission of values, traditions, and patterns of relationship has never occurred. The family, the real one made up of parents, uncles, aunts, and grandparents, has survived uncontaminated, and has persisted as something 'other' with respect to the dictatorship and its post-war implications. In Italy the elaboration of the past has not meant a radical rejection of authority, and of the authority of the family in the first place. Even the Italian 1968, when compared to the German one, was more political and less focused on the solution of psychological and relational problems.

For many years I have enjoyed this relaxed atmosphere, which has allowed me to search for my real self, to be rooted, to interweave relationships, and to face the past with less anguish. Perhaps everything really was simple. Was it perhaps only in my Protestant culture that the key could be found to explain these differences in the memory of totalitarianism, these differences that I have felt?

Some years ago, when a growing number of episodes of explicit racism began to occur in Italy too, I had to rethink it all again. I have not been so much surprised by racism in itself as I have been by the reaction of many intellectual and non-intellectual people: they use explanations, distortions, evasions in order to state that racism was something else. Why this fear? Why this anxious need to show that Italy in itself is not and cannot be racist? It is no more or less racist than other industrialized countries. But when an immigrant of colour cannot find a home, and is insulted, beaten up, and looked down on, what does it mean? Recently, when I was talking about this at a seminar of sociologists, I was warned not to 'bring out the skeletons in the cupboards'. I was suspected of 'smelling racism everywhere'. That was a difficult moment for me. I felt lost, just for a moment. But since that meeting I have tried to understand the relationships between historical memory and present sensibility better, as far as the problem of racism is concerned.

I think that it is more a question of tending to expel, to push away, rather than not see. And so I finally reach the point from which my discussion here started: *Fascist Legacy*, the BBC documentary. The first part of this film shows the horrific war crimes committed by Italians, while the second part discusses how and why they have been covered up. This film was broadcast in England in November 1989. The Italian ambassador in London officially protested to the British government. The protest did not question the truth of those documents, but the fact that only the Italian crimes were shown! A similar statement was made by Gianni De Michelis, the Italian Minister of Foreign Affairs. During the following days the Italian press reported those events. Some papers included comments in the 'world' section, others in the 'entertainment' section, and others among their *cronaca* (human-interest stories). There were also a few comments by historians in the cultural sections. That was it. Apparently, RAI 1 bought the film immediately. Whether this was to show it in Italy or in order to remove it from the Italian public is an unsolved question. However, two years have already passed.

This episode struck me. Since then I have thought a lot about explanations. History does not go back. Those generations who were young after the war and could perhaps have felt the urge to get to the bottom of such horrors, even within their own personal relationships, are not the same today. A long time has passed. Probably that one generation who by delving into that memory could have experienced a serious crisis at an individual level no longer exists. For today's youth, all that belongs to the history books. Those who still have deep ties with the past are not young. What about intellectuals? One cannot imagine that the questions of tradition, of the nation, of the Fascist past have no interest for the cultured Italian public. One only has to think of the recurrent discussion of German historical revisionism or, even more recently, of the broad discussion that the 'Jenninger case' (even this case was German) raised in the Italian press. What does it mean?

The discussion on the documentary was short and, apparently, did not raise any interest beyond the impact of mere daily news. There were some parliamentary questions concerning both the statement of the Italian ambassador in London, and the honours which had later been given to war criminals. Nothing else. It is interesting, however, to look at the arguments and the discussion that the film raised.

The central question of Ken Kirby's and Michael Palumbo's documentary was the following: in contrast to Japanese and German war criminals, why did the Italians never get tried? They were war criminals, defined as such by the United Nation Commission on War Criminals, but they were criminals for whom requests for extradition from the affected countries were completely rejected by both the American Pentagon and the British Foreign Office. In the archives there are 800 requests for extradition, advanced by Yugoslavia and Ethiopia, which have never been fulfilled. Paradoxically, the only trial, and the consequent execution of the defendant, was carried out by the British against an anti-fascist officer named Bellomo. The Allies were afraid of the socialist and, above all, of the Communist influences on Italian domestic politics. Consequently, they found the defeated Fascists somehow more reliable allies in their fight against communism as the Cold War began. Apparently there were unspoken common interests between the winners and the losers. In the short run, the Italians were interested in forming a government based on national reconciliation work (in fact, one of the first acts of the new Minister of Justice, Palmiro Togliatti, was the final closing of the trials against the fascists; and this happened as early as 1945). In the long run, without doubt, this type of politics contributed to the creation and reinforcement—on a broad level—of an Italian fascism with 'a human face'. In criticizing the delay with which Italian historiography analysed Fascist colonial politics, Professor Nicola Tranfaglia notes the long-term consequences of such politics: 'Not only did we move late; but in the last decade a vision of the Fascist regime has strongly emerged that overlooks its racism, its methods of occupation, and the aspects it had in common with National Socialism . . . In doing so, there is no doubt that we are going toward a self-proclaimed innocence that has no right to exist' (*La Repubblica*, 13 November 1989). The reconstruction of this important historical phase, as offered in the English documentary, together with the Italian reaction to the film itself, shed some light on the historical roots and the long persistence of such a vision of the Fascist regime.

'It is easy to imagine grey German bureaucrats engaged in carrying out the industrialization of genocide. How can we possibly imagine the cheerful, extroverted, pleasure-loving Italians doing the same?' This question, raised by the Sunday *Observer* when commenting on *Fascist Legacy*, evokes well-grounded stereotypes that have deeply penetrated common sense. It is uncomfortable to challenge them because these are the fundamental social representations around which public opinion has been structured. Understandably a defensive reaction and the suppression of any doubt are often the immediate responses of many people. The problem, however, takes a different shape when the speaker is a historian who has, as *Il Corriere della Sera* points out, influenced generations of high-school students: 'At least three generations of students have studied with his books, and his historical reconstructions belong to Italy's collective imagination.' Let us look at this reconstruction. I have chosen this article of *Il Corriere della Sera* of 10 November 1989 in particular, because it includes an interview with Professor Giorgio Spini and offers an exemplary and almost didactical synthesis of the main arguments of what I would call an *expulsion syndrome*.

Even the article's subtitle reveals the intention to shift the attention to the external: 'Our historians do not deny the crimes but argue with the British'. It seems to imply that the problem is elsewhere. Those crimes do not concern us: 'unfortunately they happened', as Spini says, but 'they were the responsibility of His Majesty the King, the Prime Minister Benito Mussolini, and the head of their army, the general Pietro Badoglio, and so on. These are not our responsibilities.' Spini continues: 'The British wanted us to keep the king, so they must be the ones who have problems with History.' Beside attempting, even before opening a discussion, to blame others, the resistance is evoked as a decisive shield: 'We did whatever we should and could do: we shed our blood in the battle against Fascists and Germans. We killed Mussolini and we hanged him by his feet; we overthrew our king. What do you want from us?' Finally, to end any further argument, Spini uses a sort of shopkeeper mentality, equally brutal and narrow-minded, and notably defensive: 'Our only defence is the following: the Italians during the resistance were treated just as they treated the Abyssinians and the Yugoslavs. If History was a form of national bookkeeping, one could say that their blood has paid for the blood which was shed in Dalmatia and Ethiopia.'

Another article, which appeared in the same issue of *Il Corriere*, but on its front page, explicitly rejects both the message of the film and also any invitation to discuss responsibilities. It cynically states that 'during a war it is rare to save one's soul. No army, even Her Majesty's, is spotless as far as barbarities are concerned.' Everything is simple. The incident is filed as a British problem: 'there is a feeling that they intended to force the issue against the Italians'. From being guilty, even before they could see this film, the Italians were made the victims. In this view the film was 'unhesitatingly anti-Italian'. Even *La Stampa* (11 November 1989) depicts this film as having 'strong tones', as if the horrendous images were due to the director's peculiar taste.

The left-wing press offered a different view: 'one could say that the documentary is even too good toward Italy', as *L'Unità'* remarked on 2 December 1989, when discussing the only showing of the film in Italy at the People's Festival in Florence. Furthermore *L'Unità'* enriched the debate on the film by publishing excerpts from direct testimonies of the events which it documented, already previously published in Italy. These included excerpts from chaplain Pietro Brignoli's diary, referring to the random executions in Croatia, and from the secret diary of Ciro Poggiali, who was then correspondent for *Il Corriere della Sera* in Addis Ababa. They describe the retaliations following the attempted murder of viceroy Graziani. Also *Il Manifesto* (11 November 1989) emphasized the need—even if that need comes late—for a better elaboration of historical memory: 'After a forty-five-year delay, it could happen that a people finally deals with the guilt of a part of its fathers' generation.' In the same article Lidia Campagnano mentioned the Sarajevo synagogue museum where it is possible to discover 'the Italian contribution to the extermination of the local Jewish community. One day when Brandt was chancellor he bent on his knees at Auschwitz. And he had never been a Nazi. There is no picture that portrays, for example, Signor Fanfani in tears on his knees when visiting a village rebuilt on its own ashes on the Biokovo slope in Dalmatia'.

The comments of some historians in that debate deserve particular consideration. As I have said, it was a short debate, which developed for a few days in the Italian press. Not even all newspapers dealt with it. What really strikes me in some of these excerpts is their patronizing tone and their touchiness. It sounded as if to simply state that the facts were well known to the experts was a sufficient answer to the questions the documentary raised. We have already mentioned Spini. The president of the Senate, the historian Giovanni Spadolini, 'invites us to be cautious: "We need to know the documents . . . we can raise questions, but we cannot give answers" ' (*La Repubblica*, 10 November 1989). In the same issue, the historian Renzo de Felice observed: 'It seems to me that this documentary does not add anything to what we already knew'. Dennis Mack Smith even sounded outraged: 'not at the historical reconstruction, which is highly accurate, but for its excessive and violent tone' (*L'Unità*', 11 November 1989). I am not a professional historian, but I do believe that history and historical memory need to be constantly fed by a dialectical relationship between historical data—which are known and memorized as such in archives, books and libraries—and the individual and collective elaboration of these facts in the present. I think that Nicola Tranfaglia, himself a historian, pointed out this very aspect when he criticized the ease with which one tends to expel disturbing aspects of the past from one's memory: 'to fail to recognize what Kirby and Palumbo have defined as the "fascist legacy" does not help to fully free ourselves from it. One has to accept the fact that a European fascist "model" does exist. Although it produced a variety of experiences in different countries, it had everywhere the same scorn for human beings, particularly if they were of a different colour or of a different religious or political belief. This model envisioned a territorial expansion by war and the extermination of the enemy. If we fail to recognize this model we risk removing dark chapters of our past for which the dictatorship and its supporters were primarily responsible, but that ultimately we cannot expel from our memory' (*La Repubblica*, 13 November 1989).

I am about to conclude this meditation. The episode of the British documentary has given me the opportunity to express a feeling of discomfort and anger that I have felt growing inside me for some time. These reflections have restored my calm. The gap between collective elaboration and an individual awareness is perhaps not only due, as I used to think, to the generation problem. The missing link is the elaboration of mourning. As Lida Campagnano argues: 'The Resistance has not been able to introduce to Italy what has been called the elaboration of mourning. This is primarily the recognition of an irreparable loss. In fact, the loss of brotherhood with those farmers and children who were massacred in the Biokovo village or in Cirenaica is irreparable for generations to come' (*Il Manifesto*, 11 November 1989).

I join Lida Campagnano in hoping that from this elaboration of mourning could grow an intense desire for a new sensibility towards the future.

REVIEW ARTICLES

Oral History and Italian Fascism

Alfredo Martini

In the opening lines of her book, *Torino operaia e fascismo*, (Turin Workers and Fascism) Luisa Passerini states that her original intention was to record the working-class memory of Fascism.[1] This is a useful reminder of the fact that oral history in Italy originated within labour and working-class history, and that it only gradually developed as a specific tool for the history of the working classes, as distinct from that of their institutions and organizations.

Luisa Passerini's book was a turning-point in the oral history of Fascism. Others had dealt with the Fascist era before, but this was the first systematic attempt to use oral testimony to reconstruct the way Fascism was perceived by the common people. Passerini's study, which was part of wider research on the Borgo San Paolo neighbourhood in Turin, also contributed to the inauguration of a series of innovative research projects dealing with working-class neighbourhoods and towns.

Before the 1970s (with some exceptions: see Bosio),[2] oral history was used mainly for the study of the resistance as an armed struggle and a guerrilla movement. The Borgo San Paolo project shifted the attention of oral historians toward a middle or long-range reconstruction of the relationships between a given population and its living and working environment. Also, Fascism had been seen, under the prevalent influence of the historical philosopher Benedetto Croce, as a 'historical parenthesis' in an evolutionary process toward progress and democratic liberalism; on the other hand, labour history, focusing as it did on leadership and organizations, found it difficult to deal with a historical period in which these levels were all but suppressed. The new approach seemed to make Fascism readable again: the very age of the informants and the structure of life histories made the 1920s and 1930s the crucial period.

The history of the resistance movement itself seemed to benefit from this new approach, and lengthened its time range. In a recent essay, Marco Oliva, whose *La resistenza alle porte di Torino* (The Resistance at the Gates of Turin) makes abundant use of oral testimony, notes that we need to push back 'our time range, beyond the broad but yet inadequate period of the war, 1940–45', in order to reconstruct 'a long-run period of growth, which had its roots in the 1920s and 1930s . . . and developed through the recognition of the Fascist organization of the State and the impact of Fascist propaganda, coming to fruition in the wartime experience.'[3]

The most interesting results, however, are found in research on neighbourhoods and towns. In most published essays, Fascism is looked at, above all, as a period of time, an 'epoch' in which social struggle becomes politicized and gradually turns into civil war. In the recollections of the Communists of Terni, collected and spliced together, film-like, by Alessandro Portelli, Fascism is the time in which the arrogance of factory owners and managers grows intolerable. Being a Communist is a flag which can no longer be shown at all times and in all places, and violence and repression become widespread and official. Fascism is seen in its most evident aspects, as an anti-popular system and an authoritarian regime which expands from the organization of the State to invade civil society as a whole. In the book, biographies evolve through the narrated stories, and Fascism is a chronological context, a phase of one's life which is gradually filled with meanings and moulded into a coherent design.[4]

Gribaudi on the other hand, sees Fascism as a time in which certain processes come to fruition, independent of the ideological and institutional context.[5] It is a time of modernization which induces changes in the mentality and collective culture of a social group. The thesis seems at times strained in the light of the documents and testimony used, and appears to apply mostly to individual cases; further interpretive work seems necessary.

But Fascism is also a movement: not just a context, but State, regime, and personal experience. This is apparent from Passerini's analysis of popular reactions to the regime's demographic policies. Passerini sees in this area the continuity of a popular culture which turns into resistance and widespread social opposition. Women analyse the past with most clarity, clearly detailing both official policies and violent repression, and the political and ideological mediations and flexibility imposed by the needs of everyday life.

The interviews collected by Maria Grazia Ruggerini on the history of the Santa Croce neighbourhood in Reggio Emilia insist on the inevitability of being Fascist: some Fascists were good people; others were not violent; still others are said to have joined against their will. The narrators seemed concerned not to alter a balanced collective image of the persistence of working-class and anti-Fascist identity. While they cannot claim that there were no Fascists in Santa Croce, they can at least suspend judgement or attenuate the guilt.[6] These strategies are found in all the works on the period, including research on other tight-knit and solid communities, such as Terni or the San Lorenzo neighbourhood in Rome:[7] the need to insist on a strong collective identity causes the narrators to expel or extenuate extraneous entities.

Often Fascism appears to have caused a consolidation of the collective mentality and social identity of local communities by forcing stronger ties of solidarity and unity, especially in already strong working-class groups situated in distinct productive contexts, such as Terni, the Borgo San Paolo community in Turin, and the paper-mill workers of the Liri Valley;[8] or the Roman printers studied by Lidia Piccioni.[9] Fascism itself facilitates this process by extolling certain values, such as strength or prowess, which are shared by the working-class tradition:[10] for instance, the Fascist organization of sports and leisure activities was also well received by workers.[11]

As Luisa Passerini points out, it is in the dimension of daily life that Fascism loses its totally negative connotations and appears as merely another biographic segment. Everyday life with its detailed immediate needs becomes the area of mediation, of inevitable coexistence, of little resistance, and of occasional acceptance.

In certain cases, Fascism creates events, radically altering the lives of individuals, families, and groups. Such is the tragic case of racial deportation, which manifests the most inhuman side of Fascism.[12] It is also the contradictory case of migrations to newly reclaimed lands in the Roman and Pontine plains. Franzina, Gaspari, and Martini describe in their essays the narrators' inner contrast between traditional opposition to the regime, and acceptance of the opportunities offered by migration and relocation. This is reflected in the structure of memory, the selection of events and anecdotes, the other narrative strategies which are often a mix of fine distinctions, broad omissions, and a proud recall of traditional forms of rural resistance and sabotage.[13]

Research in rural contexts does in fact draw less attention to the narrator's own perception of Fascism than to the changes that Fascism, as a regime and as a complex historical period, produced in collective mentality, and in models of social stratification.[14] This approach requires new interpretive tools, drawn in part from linguistic analysis.

Notes

1. Luisa Passerini, *Torino operaia e fascismo*, Rome, 1984 (trans. *Fascism in Popular Memory*, Cambridge, 1987).
2. Gianni Bosio, *Il trattore ad Acquanegra: Piccola e grande storia in una comunità contadina* (Tractors in Acquanegra: Little Stories and Great History in a Peasant Community), Bari, 1981.
3. Gianni Oliva, *La resistenza alle Porte di Torino* (The Resistance at the Gates of Turin), Milan, 1989; id., 'Notes on a local history of the resistance', *Studi piacentini*, 7 (1990), 211–20.
4. Alessandro Portelli, *Biografia di una città: Storia a racconto Terni 1930–1955* (Biography of a City: History and Narratives of Terni 1830–1955), Turin, 1985.
5. Maurizio Gribaudi, *Mondo operaio e mito operaio. Spazi e percorsi sociali a Torino nel primo novecento* (Working-Class World and Working-Class Myth: Social Spaces and Pathways in Turin in the First Half of the Twentieth Century), Turin, 1985.
6. Antonio Canovi, 'Un posto da abitare' (A place to live), in id., M. Mietto, and M. G. Ruggerini, *Nascita di una città. Il territorio di Santa Croce: la storia, la memoria, le 'Reggiane'* (Birth of a City. The Santa Croce Neighbourhood: History, Memory, and the Inhabitants of Reggio Emilia), Milan, 1990, 35–83; Maria Grazia Ruggerini, 'La vita di tutti i giorni' (Daily life), ibid. 139–214.
7. Lidia Piccioni, *San Lorenzo: Un quartiere romano durante il fascismo* (San Lorenzo: A District in Rome under Fascism), Rome, 1984.
8. Alfredo Martini, 'Essere operai: L'uso delle fonti orali per la storia di un gruppo sociale' (Being workers: using oral sources for the history of a social group), *Rassegna degli Archivi di Stato*, 18/1–7 (1988), 265–75.
9. Piccioni, ' "Il tipografico é una bestiaccia che non si assoggetta a nessuna disciplina": Storia, memoria e immaginario dei tipografi romani (1927–40)' (Printing is a brute which no discipline can subdue: History memory, and fantasies of Roman printers, 1927–40), in D. Sacchi, G. Sircana, L. Piccioni, and A. Lombardo, *Operai tipografi a Roma 1870–1970* (Printing-Workers in Rome, 1870–1970), Milan, 1984, 279–396.
10. Portelli, *Biografia di una città*.
11. Canovi, 'Un pasto da abitare'; Gribaudi, *Mondo operaio e mito operaio*.
12. Anna Bravo and Daniele Jalla (eds.), *La vita offesa: Storia e memoria dei lager nazisti nei racconti di duecento sopravvissuti* (Outrage to Life: History and Memory of the Nazi Concentration Camps in the Narratives of Two Hundred Survivors), Milan, 1986.
13. Emilio Franzina, 'Una emigrazione nazional-popolare: I coloni veneti nell'Agro Pontino' (A national migration: Settlers from the Veneto in the Agro Pontino), in id. and A. Parisella (eds.), *La 'Merica' in Piscinara: Emigrazione, bonifiche a colonizzazione veneta nell'Agro Romano e Pontino tra fascismo e post-fascismo* ('Merica' in Piscinara: Migration, Land Reclamation, and Settlement from the Veneto in Agro Romano and Agro Pontino Under Fascism, and in the Post-Fascist Era), Abano Terme, 1986, 31–119; Martini, 'Coloni e braccianti veneti nell'Agro Romano: Maccarese negli anni Trenta' (Settlers and farmworkers from the Veneto in the Agro Romano:

Maccarese in the 1930s), ibid. 131–60; Oscar Gaspari, 'La "Merica" in Piscinara: I veneti pontini dalla colonizzazione fascista agli anni Sessanta' ('Merica' in Piscinara: the Pontines of the Veneto from the Fascist colonization to the 1960s), ibid. 217–89.

14. Piero Brunello, 'Contadini e repetini: Modeli di stratificazione' (Peasants and independent labourers: models of social stratification), in S. Lanaro (ed.), *Storia d'Italia. Le Regioni dall'Unità ad Oggi: Il Veneto* (History of Italy. The Regions from the Unification to Today: The Veneto), Turin, 1984, 859–909; and Martini, 'Colonie braccianti'.

Stories of Everyday Life in Vichy France

Dominique Veillon and *Danièle Voldman*

Up to now, French historians have been little inclined to study France between 1940 and 1944 from oral sources, which have been thought to be too subjective. There are several reasons for this. In the first place, the subject matter, oriented especially toward political and institutional aspects of Vichy France, has been analysed through written public records which satisfactorily fulfil their purpose. Secondly, a low value has been attributed to everyday life by historiography in the contemporary period. Finally, personal statements difficult enough for historians to handle in 'normal' times, entail complicated risks when it comes to studying aspects of the 'Franco–French war'. Academic historians argue that the written record legitimizes validity, and that oral accounts remain subject to distortion: they are only just beginning to win professional acceptance. Even very recently, at an international colloquium on Vichy, which took place in Paris in June 1990, it was noticeable that the life of French people, in its daily banality, had been studied too little.[1] In contrast with the Anglo-Saxon countries, where the challenge has long since been recognized, in France, time has been lost, and is only now beginning to be made up for.

Paradoxically, therefore, it is Anglo-Saxon historians who have most studied the day-to-day life of occupied France. Research by Sarah Fischman in 1986 on the lives of French prisoners-of-war's wives demonstrated how much interest can be generated by a well-conducted oral survey.[2] Aspects, which until then had been passed over in silence, were brought to light, such as child-rearing, sexual problems, professional life, and the material difficulties experienced by these women who had become heads of families by necessity. In the same way, the interviews conducted by Paula Schwartz,[3] who studied Communist women in the resistance, enabled the domain reserved for women, and the difficulties of their lives during this period, to be more closely identified. Many women played so-called secondary roles (liaison officer, secretary) as *résistantes*, as well as continuing to be mother, spouse, and so on. Without these women's actions and effective help, the resistance would never have made its name. This aspect of their double life is clearly revealed in the book by Stéphane Coutois, Adam Rayski, and Denis Peschanski,[4] who, in addition to unpublished records, used about thirty oral accounts.

Only one area has been genuinely explored with the help of life stories, but as this focuses on daily life in dictatorial or totalitarian regimes, it presents two major and exceptional characteristics: on the one hand, stories of deportees in the Nazi extermination camps; and on the other, the work of sociologists who are as much concerned with providing an account of this period as with helping to restore the identity of their subjects. By telling their stories, these people have relived, several decades later, their extreme experiences.[5]

Nevertheless, if historians have demonstrated a lack of enthusiasm for the 'everyday' and the 'word of mouth', several advances have been made in recent years. Researchers, more and more conscious of the need to question witnesses about their activities in the resistance, and of keeping a record, are systematically recording the interviews that they conduct. Laurent Douzou in 'Libération-Sud' and Olivier Wieviorka in 'Défense de la France' have, in their respective dissertations, systematically integrated an important body of statements.[6]

Some of these recordings appear to encompass not only the period of the war itself, but the whole of a life story, in order to put this critical period into context. A not insubstantial part of the story is given over to daily life. Besides allowing us to grasp the rhythm of the times, by not focusing attention on isolated, out-of-the-ordinary moments, this method results in an accumulation of the minutiae of existence. Thus the exceptional is rooted in normality. In this historical quest, the domestic, the concrete, and the individual are single links in a chain which connects stories to history.

These studies allow contemporary history to rejoin one of the important streams of the École Historique Française which in earlier periods explored the attitudes, the private life, and the individual destiny of the man and woman in the street. Thus 'word of mouth' takes on an incalculable and uniquely precious dimension, as soon as it is a matter of stories of everyday life. The war and politics become the stage on which the existence of witnesses plays itself out. This was noticeable recently at the colloquium organized by the University of Nice–Sophia Antipolis in December 1990 on twentieth-century wars in oral accounts. From recordings gathered within the framework of the Université du Troisième Age, of Caen,[7] with around thirty witnesses, a paper was presented on the Normandy landing as seen by the civil population.[8] In addition to the important studies of the Allies' strategy regarding the landing, the interest of the work rests on the recapturing of the way of everyday life of those who were destined to be liberated, which had long since been neglected as a historical subject. We know about the impact of the bombardments on the Normandy population, and the traumas they suffered, but how they lived through the long weeks of the battle, in concrete terms, is less well known. In Caen, for example, the life of the population, taking refuge in the quarries, in the churches, or hidden in the surrounding countryside, is now more familiar.

In the collection of accounts, domestic chronology prevails over national chronology. The famous D-Day, 6 June 1944, was, for those who were not soldiers, the day of the *baccalauréat* examination, the day of a religious ceremony, or the day of a parent's birthday. This private event becomes the temporal focus of the memory, a necessary reference point. The oral survey thus brings to light a local memory out of line with the national memory.

Another kind of research is that which concerns particular age-groups or socio-economic classes. One could cite the film-makers whose working conditions were changed between 1940 and 1944; or the Zazous, those youths who under the occupation and the Vichy regime demonstrated their rejection of society and the adult world by their choice of clothing and their taste for American music; or the architects battling with the corporate organization of their profession and with German pressures in the building and construction industry; or the factory workers in the dressmaking trade who suffered from the most banal aspects of the war, that is the restrictions and shortages.[9]

Thus, thanks to a renewal of historical subjects and of traditional sources, a more refined landscape of France during these times is gradually being delineated. The history of the expelled is being extended from the *Résistants* and the Jews, to the whole of the imprisoned population. It is the subject of a vast survey which is starting at the Institut d'Histoire du Temps Présent in Paris encompassing everyday life, social life, and the history of nutrition. New topics and new research imply the invention of new sources. It is here that life stories should come into their own.

Notes

1. Jean-Pierre Azéma, François Bédarida, Denis Peschanski, and Henry Rousso (eds.), *Le Régime de Vichy et les Français* (The Vichy Regime and the French), Paris, 1991.

2. Sarah Fischman, *Divided by War: Prisoner of War Wives in France, 1940–45*, New Haven (Conn.), 1991.
3. Paula Schwartz, 'Partisans and gender politics in Vichy France', *French Historical Studies*, 16/1 (1989), 126–51.
4. Stéphane Courtois, Denis Peschanski, and Adam Rayski, *Le Sang de l'étranger: Les Immigrés de la MOI dans la résistance* (Blood of the Stranger. Immigrants of the Industrial Workers' Group in the Resistance), Paris, 1989.
5. See esp. Nicole Lapierre, *Le Silence de la mémoire: A la recherche des Juifs de Plock* (The Silence of Memory: In Search of the Jews of Plock), Paris, 1989; and Michael Pollak, *L'Expérience concentrationnaire* (The Experience of the Concentration Camp), Paris, 1990.
6. Laurent Douzo, Ph.D. thesis on the movement and the journal *Libération-Sud* (forthcoming); Olivier Wieviorka, Ph.D. thesis on *Défense de la France* (forthcoming); interviews will be deposited in the Archives Nationales, Paris.
7. Étienne Fouilloux and Dominique Veillon, 'Mémoires du débarquement en Normandie' (Memories of the Normandy Landings), in François Bedarida (ed.), *Normandie 1944. Du débarquement à la libération* (Normandy 1944: From Landings to Liberation), Paris, 1987, 214–27.
8. Veillon, 'La débarquement de Normandie vue par la population civile', in *Les Guerres du XXème siècle à travers les témoignages oraux* (Actes du Colloque de Nice, 1991).
9. Pierre Bertin Maghit, *Le Cinéma sous l'occupation: Le monde du cinéma français de 1940 à 1944* (Cinema during the Occupation: The World of French Cinema, 1940–4), Paris, 1989; Emanuelle Rioux, 'Les Zazous, un phénomène socio-culturel pendant l'occupation' (The Zazous: a socio-cultural phenomenon of the occupation), mémoire de maîtrise, Université de Paris X—Nanterre, 1987; Danièle Voldman and Frédérique Boucher, *Les Architectes sous l'occupation* (Architects in the Occupation), 1992, forthcoming; Veillon, 'La moda en tiempos de guerra' (Fashion in wartime), *Historia y fuente oral*, 3 (1990), 105–21.

Voices from the Choir: Reflections on the Development of Oral History in Russia

Irina Sherbakova

Although many of those now working with oral history in Russia feel themselves pioneers, it certainly cannot be said that they have begun with a fresh start from nothing. Russian oral history draws on quite extensive traditions, and without going back into grey antiquity, we could instance Pushkin's *Ístoria Pugachevskovo bunta*,[1] which made use of witnesses of the eighteenth-century peasant uprising. At the turn of this century, there was a considerable scholarly discussion of oral sources such as legends, sagas, and tales.

In the 1920s, however, after the shocks of the revolutions and civil wars, the use of oral sources gained a new significance. The use of direct oral witness was given much more priority in literature and in journalism. This was partly because the revolutionary wave brought large numbers of illiterate or semi-literate people to the social surface: people, who could only record and convey the memory of their experience orally. But simultaneously, and paradoxically, with the intensification of a totalitarian ideology requiring a unitary form of thinking, any serious and reliable historical investigation of the revolution and civil wars became impossible, and in this situation oral sources were manipulated to give the illusion of authenticity to historical falsification. The maintenance of Stalinist credibility eventually became a game with history, continually presenting new façades, like a children's game with cubes, until eventually its own foundations were undermined. Absurd falsifications were supposed to be canonized into an eternal truth.

Thus during the 1930s a number of books were published to record recent history with the help of misleading oral accounts. For example, *Kanal imeny Stalina*,[2] a joint work by twenty famous Russian writers on the construction of the White Sea–Baltic Canal by prison labour, is a book in praise of slave labour in the name of Socialism. Prisoners' witness was used in it to lie about the true character of the recently created Gulag. An equally false, if less odious book, is *Kak mi stroilly metro*.[3] The solemn inaugural journey of the Moscow metro took place on 14 May 1935, and on 21 June a luxurious second volume, with a gold-lettered jacket, was already published *Istorii metro*.[4] There is—understandably—not the slightest reflection in it of the monstrous exploitation of manual labour which had enabled the metro to be built with such incredible speed.

Thus on the one hand, 'official' oral sources were used during this period for the distortion of history, while on the other hand, the true story of the Stalinist repressions, the prisons, and the Gulag only came to survive and be transmitted in oral form, because people were so afraid. Therefore, when after Stalin's death former convicts began to return from the camps, and remembering became less dangerous, the first information seeped out by word of mouth. Some people even 'performed' with their friends as story-tellers, about the Gulag. When in the late 1950s Alexander Solzhenitsyn set himself the aim of writing the history of the Gulag, he basically used oral accounts: and this was probably the first instance in Soviet history of their use for the recovery of the truth.

However, at that time, when he had completed *Archipelago Gulag*, Solzhenitsyn was only able to publish it abroad.[5] Only books of oral stories about war were managing to get through the censorship. Under Brezhnev, the subject of the Great Patriotic War of 1941–5 was, in contrast to the theme of Stalinist repression, not only permitted but positively encouraged, even if with the intention of a deliberately unbalanced over-varnished 'heroic' picture. Nevertheless, under this theme a number of relatively truthful books, based on oral witness, were published. These were important because for the first time they showed the fate of ordinary people in the war, and also revealed some previously hushed-up episodes. For example, the Belorussian author Adamovicha published accounts of people who were miraculously rescued from villages burnt down by the SS during the German occupation of White Russia.[6] This was followed by a joint work with Daniel Granin, *Blockadnaya kniga*,[7] which included the testimonies of people who had survived the Leningrad blockade of 1941–4, when the German armies had encircled the city, and hundreds of thousands of people had died of starvation.

It was also in the 1970s that television began recording oral testimony: for example, the writer Konstantin Simonov presented a series of programmes, 'Shol soldat' (The soldier set out), in which he conversed with ex-soldiers about their wartime memories.

Probably the most interesting books of this phase, from the point of view of oral history, are those of the writer and journalist Svetlana Alexievicha. In *U voiny ne jenskoe litso*,[8] she brought together the stories of women at the front during the 1941–4 war: altogether 800,000 women served in different military capacities in the war. This was a particularly new and valuable perspective because hitherto all accounts of the war had been essentially from a male standpoint. Women's memories extended much more to human feelings about the war. Her next book, *Posledniye cvidetely*, is based on the stories of those who were children in the war, and so the youngest generation who remember it.

It is no accident that with the beginning of perestroika in 1985–7, the oral testimonies of survivors literally gushed forth in the mass media: they were the most immediately accessible means for the recovery of historical truth about decades of totalitarianism. The process began with the victims of the Gulag, but then widened to include the questioning of witnesses of collectivization and famine in the early 1930s. Oral evidence was also brought to bear on the most recent past, for example, to investigate the truth about the war in Afghanistan. Alexievicha, once again, has published on this: her book *Svinstovye mal'chiky*,[10] serialized in 1990 in *Drushda Narodov*, uses the recollections of participants—and their relatives—in the war.

Gradually, groups of oral-history enthusiasts have appeared, such as the Soviet society Memorial (not just in Moscow but also in other cities and republics), which is playing an important role in recording the victims of Stalinism. One group is at the Moscow State Institute of History and Archives. They have been interviewing witnesses of the 1930s' famine in the villages, and representatives of some of the national groups who suffered mass deportation under Stalin's repression. And more slowly, sound archives have also begun to develop.

There is no doubt that there are truly boundless possibilities for oral history in Russia. And now that we can at last speak and write of all that was kept silent for so many decades, and people are gradually losing their cumulatively stored-up fear, the oral historians' task will be not only to collect historical evidence about our still often unknown past, but to help a people who have survived life in unfree conditions to see the true value of biography and the importance of individual perceptions of historical events.

Notes

1. Alexander Pushkin, *Istoria Pugachevskovo bunta* (History of the Pugachev Rebellion), St Petersburg, 1834.
2. *Kanal imeny Stalina* (The Stalin Canal), Moscow, 1935.
3. *Kak mi stroilly metro* (How we Built the Metro), Moscow, 1935.
4. *Istorii metro* (The History of the Metro), Moscow, 1933–4.
5. Alexander Solzhenitsyn, *Archipelago Gulag* (The Gulag Archipelago), Paris, 1973.
6. Alexia Adamovicha, *Ya iz ognennoy derevny* (I am from the Fire Village), Minsk, 1977.
7. Id. and Daniel Granin, *Blockadnaya kniga* (The Blockade Book), Moscow, 1982.
8. Svetlana Alexievicha, *U voiny ne jenskoe litso* (War does not have a Woman's Face), Minsk, 1985.
9. *Posledniye cvidetely* (The Last Witnesses), Moscow, 1985.
10. *Svinstovye malchiky* (Boys of Lead).

Aspects of Recent Oral History in Germany

Alexander von Plato

Oral history was not highly regarded in the universities of East and West Germany until recently. But in the last few years a valuable body of historical literature has been published, partly based on oral sources, though defining itself not as 'oral history' but as 'history from below': the history of everyday life or the history of experience. In 1978, comparatively late in the European context, Lutz Niethammer published an article about oral history in the USA, thus promoting oral history in the German historical sciences. Meanwhile, outside the universities, a number of forums were established, such as the History Movement with its History Workshops, founded at the end of the 1970s, and now publishing the journal *Geschichtswerkstatt* (History Workshop).

The West Berlin History Workshop has up till now published seven volumes, a colourful mosaic dipping into the history of Berlin. The third volume, for example—a key volume—deals with what was probably the most well-known working-class district, the Wedding—though not only the Red Wedding of the 1930s celebrated by Brecht but, as the title makes clear, life in the Wedding in post-war Berlin. Life stories vividly demonstrate the impact of the partition of the former capital on its citizens and their families.[1] Other volumes in the series include one concerned with another district of Berlin, the Lindenhof, a co-operative settlement; and one published in 1988, showing the importance that the respective ruling powers attributed to street names for their ideological and political influence. Another volume contains one single life story: the memories of a Jewish judge and lawyer who lost her job in 1933, and emigrated together with her husband to Palestine, before returning to Berlin in 1953, where she became a lawyer and notary, having been refused permission by her profession to be a judge. It is based on conversations and on her personal notes. All these books are distinguished by their professional layout, and contain numerous photographs, leaflets, and other documents, which supplement the texts.

Another forum outside the universities was the trade unions with their working group History from Below, bringing together about ninety local-history trade-union working groups.[2] Yet another forum is very specifically German: the Federal President's award Schülerwettbewerb Deutsche Geschichte (German History Competition for Schools).[3] Schoolchildren were, and still are, asked to research particular fields of German history in their immediate environment, and the best projects receive prizes. One focus of these competitions, among others, had been National Socialism and the war. Tens of thousands of local projects have been undertaken, involving thousands of witnesses, and numerous important studies and documents have been produced. Another forum where professional historians, together with participants in their courses, have been able to work on local-history themes—including interviewing witnesses—has been the adult evening classes. The best-known example of this work is the *Hochlarmarker Lesbuch* (Hochlarmarker Reader), a richly illustrated book describing the history of a miners' district from the point of view of its inhabitants, for which the research also gave rise to many exhibitions and lectures.[4]

At the universities, oral sources have been mainly used by ethnologists[5] and sociologists,[6] but historians have also produced a number of studies using, among others, oral sources,[7] some of which focus on how people came to terms with National Socialism in the post-war period. In view of the fact that many of today's most influential people lived their youth under National Socialism, it is hardly surprising that, especially recently, many books on youth under National Socalism, during the war and in the post-war period, have been written.[8] Other characteristic oral history themes which have been studied include anti-Semitism, minorities, and women. All of these studies point to the fact that there is hardly anything that could be called 'pure' oral history research in the (old) Federal Republic: rather, there are a number of historical studies of everyday life, based on personal experience, which use oral as well as other 'subjective' sources, for instance diaries, letters, and photo albums. These studies demonstrate the high degree of interdisciplinary co-operation in research on personal experience, which is also apparent in the only relevant journal.[9]

After the *Wende* (Transformation) in 1989, Germany became one of the few countries where people had experienced at least three, or even five, political systems, and were therefore particularly interesting as subjects for oral history studies. Since 1989, a number of such studies have focused on German Democratic Republic society before the

Wende. Here, too, international oral-history themes have played an important part, notably in studies by Robin Ostow,[10] and Wolfgang Herzberg.[11]

In contrast to Ostow's work, which is based on notes made during or after her interviews, Herzberg, who grew up in the GDR, and still lives there, presents six very extensive life stories, which have been edited, and are not anonymous. In 1985, Herzberg published one of the first GDR oral history studies arising from interviews with workers at the Narwa electric-bulb factory. Both Ostow's and Herzberg's studies are impressive, not only in their treatment of central questions about the Holocaust and the subsequent reactions of the German population to it, and the problems of survivors and of those born later, but also in addressing the problematical relationship of the early, but in the end transient, anti-fascist movement in the GDR, which had tried to confine the industrial assassination of the Jews to a category of political resistance.

The first book about homosexuals in the GDR, by Jürgen Lemke, published both in the West and in the East in 1989, should also be mentioned. Lemke himself was only just starting out on the career ladder in a ministry in the Democratic Republic when he came under surveillance by the Stasi (internal security) because he shared his life with another man. Talking about what had become his now obvious problems as a gay man in the GDR led to further conversations, and later to a book of interviews, with a preface by Irene Runge.[12]

Women in the GDR after the *Wende* are the subject of a book by Erica Fischer and Petra Lux. Fischer is an Austrian journalist and activist in the Austrian feminist movement; Lux is also a journalist, but from the GDR. She was a leader of the Neues Forum (Opposition movement) in Leipzig. They each wrote their own introduction, but the discrepancies which might have been anticipated are not evident in the book, which consists mainly of interviews adapted for publication.[13]

Finally, I would like to mention one of the books published after the *Wende*, by Clement Wroblewsky. It consists of interviews exclusively with old friends and acquaintances, mainly intellectuals. The conversations took place in the first few months after the *Wende*, and are still deeply influenced by the high hopes of that period, which, so soon, already seem to have vanished. But because attitudes to reunification apparently changed so rapidly, the book has almost at once become a historical document: only the future will decide whether the hope or the disillusion will endure.[14]

Notes

1. *Der Wedding—hart an der Grenze: Weiterleben in Berlin nach dem Krieg* (The Wedding—Hardship at the Frontier: Getting on with Life in Berlin after the War), Berliner Geschichtswerkstatt, 3; Berlin, 1987.
2. Manfred Scharrer, *Macht Geschichte von Unten: Handbuch für gewerkschaftliche Geschichte vor Ort* (Making History from Below. Handbook for Trade-Union Local History), Cologne, 1988.
3. Dieter Galinski and Ulla Lachauer (eds.), *Alltag in Nationalsozialismus 1930 bis 1939: Jahrbuch zum Schülerwettbewerb Deutsche Geschichte um den Preis des Bundespräsidenten* (Daily Life under National Socialism 1930–9: Yearbook of the School Competition in German History for the Federal President's Prize), Brunswick, 1982.
4. *Hochlarmarker Lesebuch. Kohle war nicht alles: 190 Jahre Ruhrgebietsgeschichte* (The Hochlarmark Reader. Coal was not Everything: 190 Years of History of the Ruhr District), Oberhausen, 1981.
5. e.g. Albrécht Lehmann, 'Autobiographische Erlebungen in den sozialen Unterschichten: Gedanken zu einer Methode der empirischen Forschung' (Autobiographical testimony in social history: Thoughts on a method of empirical research), *Zeitschrift für Volkskunde*, 73 (1977), 161–80.
6. e.g. Peter Alheit, Wolfran Fischer-Rosenthal, and Erika M. Hoerning, *Biographieforschung. Eine Zwischenbilanz in der deutschen Sociologie: Werkstattberichte des Forschungsschwerpunkts Arbeit und Bildung* (Biographical Research. An Interim Statement for German Sociology: Working Notes on Critical Points of Structure and Organization of Work in this Field), Bremen, 1990.
7. Alexander von Plato, 'Oral History als Erfahrungswissenschaft. Zum Stand der mündlichen Geschichte in Deutschland'. (The practice of oral history: on the status of oral history in Germany), *BIOS: Zeitschrift für Biographieforschung und Oral History*.
8. e.g. Sibylle Meyer and Eva Schulzer, *Auswirkungen des II. Weltkrieges auf Familien. Zum Wandel der Familie in Deutschland* (Consequences of the Second World War for Families: Change in the Family in Germany), Berlin, 1989; Lutz Niethammer (ed.), '*Die Jahre weiss man nicht, wo man die heute hinsetzen soll': Faschismuserfahrungen im Ruhrgebiet* ('The years when no one knew where they would be from one day to another': Telling about Fascism in the Ruhr), Lebensgeschichte und Sozialkultur im Ruhrgebiet, 1; Berlin and Bonn, 1983) id. (ed.), 'Hinterher weiss man, dass es richtig war, dass es schief gegangen ist': Nachkriegserfahrungen im Ruhrgebiet ('Afterwards we knew that it was true, that it had gone off the rails': Post-War Testimony in the Ruhr District), Lebensgeschichte und Sozialkultur im Ruhrgebiet, 2; Berlin and Bonn, 1983; id. and von Plato (eds.), '*Wir kriegen jetz andere Zeiten': Auf der Suche nach der Erfahrung des Volches in nachfaschistichen Ländern* ('Now we'll get to other times': The Search for Popular Testimony in Post-Fascist Regions), Lebengeschichte und Sozialkultur im Ruhrgebiet, 3; Berlin and Boon, 1984; Gabriele Rosenthal (ed.), *Die Hitlerjugend Generation: Biographische Thematisierung als Vergangenheitskewältigung* (The Hilter-Youth Generation: Biographical Themes as Past Accomplishments), Essen, 1986.
9. *BIOS: Zeitschrift für Biographieforschung und Oral History*, ed. by Werner Fuchs-Heinritz, Charlotte Heinritz, Albrecht Lehmann, Lutz Niethammer, and Alexander von Plato, sociologist, educationist, ethnologist, and historians respectively.
10. Robin Ostow, *Jüdischen Leben in der DDR* (Jewish Lives in the GDR), Frankfurt-on-Main, 1988.
11. Wolfgang Herzberg, *So war es* (That's How It Was), Halle and Leipzig, 1985.

12. Jürgen Lemke, *Ganz normal anders: Auskünfte schwuler Männer aus der DDR* (In Other Ways, Completely Normal: Interviews with Gay Men in the GDR), Berlin, Weimar, and Frankfurt-on-Main, 1989.
13. Erica Fischer and Petra Lux, *Ohne uns ist kein Staat zu machen: DRR-Frauen nach der Wende* (Without Us There Would Have Been No State: Women in the GDR After the Transformation), Cologne, 1990.
14. Clement Wroblewsky (ed.), '*Da wachst Du eines Morgens Uff und hast 'nen Bundeskanzler': Wie DDR-Bürger über ihre Zukunft denken* ('So you wake up one morning and there's no Chancellor': How Citizens of the GDR Think about their Future), Hamburg, 1990.

BOOK REVIEWS

Michael Pollak, *L'Expérience concentrationnaire: Essai sur le maintien de l'identité sociale* (The Concentration Camp Experience: An Essay on the Maintenance of Social Identity). Paris: Métailié, 1990.

This book is based on the assumption that every extreme experience reveals the form and the conditions of a normal experience, the very familiarity of which makes analysis deceptive. In this research, the experience of one particular concentration camp, the Auschwitz-Birkenau extermination camp, is taken as a touchstone to reveal identity as self-image, both for oneself, and for others. The exceptional character of such an experience makes the two phenomena of identity and memory more visible, and therefore analysable. At the same time, Michael Pollak reminds us that to construct the object of history has been, in this case especially, difficult work, Such a construction could be defined as the passage from the sources—archival documents, including judicial testimonies and testimonies collected by historical commissions, interviews, and observations—to the presentation of those relationships which are considered particularly significant between the characteristics of the deported people, the functioning of the camp, the ways of adaptation to that universe, and the readaptation to ordinary life after liberation. Usually, judicial documents do not give much information about relationships among detainees, while historical testimonies do. Pollak also uses a number of published autobiographical accounts, like the well-known ones by Louise Alcan, Edith Bruck, and Suzanne Birnbaum, as well as two unpublished manuscripts. It is interesting to notice that more than one-third of these accounts were written before 1949, a sign of the will to overcome and transmit an inconceivable past.

A central part of the book comprises the reconstruction of the biographies of three women: Margareta from Vienna, Ruth from Berlin, and Myriam from Paris. The three life stories are chosen from a corpus of twenty. The experiences recalled in them show that there were very different ways of living in a concentration camp: from the languages one spoke, and the capacity for communication in a multinational universe, to the ways one chose to adapt oneself. It is important to remember at this point that altogether less than two per cent of the survivors from concentration camps have been able to leave a testimony of their experience. Among them, testimonies by people who were involved in the functioning of the camps are very rare, and yet the exercise of 'privileged' positions increased the chances of survival. One striking example of the problems connected with such memories is given in this book by the story of Ruth. She had told her interviewers that deported people working as doctors and nurses were obliged to kill Jewish new-born babies in order to save their mothers, who risked being sent to the gas chambers with their babies. She had spoken not only as an eyewitness, but as an actor who found herself under extreme pressure in those circumstances. When reading the transcript of her interview, however, she changed the passage in such a way as to take up the role of somebody who had only heard of, but never seen such things actually being done. In the discussion that followed, Ruth insisted that the second version was the true one, 'since so many things mix in my memories of more than forty years ago that sometimes I cannot distinguish between what I have lived through, and what I have read or heard from somebody else'.

An important observation on the structure of these memories is that in all the versions about the same life one can distinguish a hard centre. This central core can be found every time, identical, word for word, and it even shapes the form of orality to the extent that exactly the same tone is used for certain phrases. At the same time invariance such as the one indicated by such a phenomenon is coupled with its very opposite: Pollak indicates that frontiers between the unsaid on the one hand, and the repressed and forgotten on the other, are constantly being moved. These two observations together point to the dialectical and constantly moving nature of memory.

Of particular interest is the section where the use of personal pronouns is analysed, following the suggestions of Benveniste, in order to shed light on some psychological phenomenon—for example, talking to oneself in the second person, or talking about oneself in the third person, as a result of changes in the psychic structure of the people who lived through the *expérience concentrationnaire*. In this respect, a proposal by Pollak would be worthwhile debating by historians: he believes that two opposite interpretations, such as the ones by Bettelheim and des Prés, could be treated as complementary on the basis of historical sources, whilst appearing irreconcilable on the grounds of theoretical oppositions between psychoanalysis and sociobiology. What Bettelheim considered the main value of our civilization, the search for individual and collective dignity that allowed people to survive, expresses, according to des Près, a will to dominate nature and others that leads eventually to the self-destruction of the species. On the contrary, des Près sees in one quality of the survivors, i.e. their capacity to distance themselves, and to be sceptical about current morality, the sign of a new morality, practical and modest, oriented towards the survival of the species, and breaking the traditions of the great values of our civilization. Pollak claims that, in actual experience, these theories represent the two poles of experience in concentration camps, between which a whole series of intermediate positions can be found, depending on the resources of the individual and the social links which give form to his/her identity.

The intertwining of empirical data, new sources, participating interpretation, and theoretical insights make this book valuable not only for oral historians, but for historians in general, as well as for all readers.

Luisa Passerini

Alessandro Portelli, *The Death of Luigi Trastulli and Other Stories: Form and Meaning in Oral History*. Albany: State University of New York Press, 1991.

Alessandro Portelli's *The Death of Luigi Trastulli and Other Stories*, a part of the SUNY Series in Oral and Public History under the editorship of Michael Frisch, is a collection of essays which have appeared previously, although not always in English, and in somewhat different forms. The essays are on a variety of subjects in both American and Italian contexts. All are well written and important for the specific insights they provide, and for the dimensions they give to the role of oral history. Cumulatively, they have an additional identity in that they provide an umbrella of understanding for the complexity, diversity, richness, and promise of oral history. In addition, they make the examination of oral history practice and experience in an international context possible.

The book is artistically and creatively arranged, with at its heart a section on 'Two Industrial Cultures': Terni in Italy and Harlan County, Kentucky, in the USA. Although Portelli cautions us against taking a purely comparative view of the two cultures, since the studies are in different stages—the former complete; and the latter still in process—it is suggestive of his approach that he studies two working-class areas. The emphasis in all his essays is upon fieldwork, the relationship of narrator to interviewer, and the importance of oral testimony as a literary exercise—which means that transcripts of the spoken word run the risk of editing, which alters the literary character. From the literary character of oral history, Portelli proceeds to an analysis that includes studies of the misrepresentation of chronology and the fabrication of events. By dealing with these factual errors from the standpoint of symbolism, mythology, and literary device, rather than as mistakes of historical fact, he arrives at compelling characterizations of the nature of individuals and communities—their aspirations, hopes, and dreams. As he argues, 'The discrepancy between fact and memory ultimately enhances the value of oral sources as historical documents. It is not caused by faulty recollections . . . but actively and creatively generated by memory and imagination in an effort to make sense of crucial events and of history in general' (p. 26).

Readers will undoubtedly find some essays of more interest than others, according to personal experiences and intellectual outlook. But what is particularly impressive about this book is that the essays, written over a long period of time, display a progression of ideas in a co-ordinated order that yields a real theoretical synthesis of oral history. This may indeed be a book of essays, but their interconnection almost belies that fact.

Carl Ryant

Marcella Filippa, *Avrei capovolto le montagne: Giorgina Levi in Bolivia, 1939–46* (I Could Have Scaled the Mountains: Georgina Levi in Bolivia, 1939–46). Florence: Giunti, 1990.

This book is the story of a woman creating a life in political exile, in a peripheral context of the Jewish diaspora, in a dramatic confrontation with the Third World. Giorgina Arian Levi, raised in the anti-Fascist environment of the Jewish middle class in Turin, is forced to leave her country with her husband, a German Jew exiled from Hitler's Germany, after Mussolini's 1938 legislation discriminating against Italian Jews. The world was closing around people like them: they were to spend their seven-year exile in a country which they had barely heard of before—Bolivia.

'Half my soul belongs to Bolivia, that is where I became what I am . . . as a Communist, as a woman, as a wife, in a way also as a teacher. It was an experience in which I learned to live, not to be afraid of death, to face dangers, to realize that I could face them without fear.' In the villages of the Andes, even in La Paz, her life was suddenly reduced to the barest essentials. Travel, personal hygiene, and housekeeping were all less than rudimentary, and often dangerous: as she taught Indio children the use of the handkerchief—much as Booker T. Washington used to teach newly freed American Blacks the use of the toothbrush—as a civilizing experience, she also seemed to be holding on to at least a rag of her European and middle-class upbringing. There was no such thing as respectability: she moved into a prostitute's house, the only conceivably human lodgings in the village; and when her first stint as a teacher of Latin in the university met with unexpected antagonism, she defiantly took a job as an apprentice hairdresser. Emotions were hardened by the constant difficulty and deaths: even the loss of her baby in childbirth was virtually taken in her stride. Doing became more important than feeling, perhaps the only possible way of feeling, and Levi developed into a 'woman of action'.

Much space is devoted to describing the various groups of exiled Jews who had sought temporary refuge in Bolivia. They came from different parts of the world, and would later disperse to places as far away as China, and as near as her own home town of Turin. Yet, their community is described as often narrow-minded and ridden with prejudices of nation and class. 'I think emigration breaks down the communal element in Jews. It divides them': yet, as always, she turns the negative discovery into a positive lesson, against stereotypes. 'I discovered', she concludes, 'that Jews are human beings who struggle to survive, who do not devote themselves exclusively to helping their brethren in distress, who, in other words, are men like all others, with their weaknesses and virtues.' After she left Bolivia, the sense of her own Jewishness was keener than ever.

Throughout, Levi sees herself as little more than a sojourner: Bolivia became part of her, but she was never part of it. She was growing more intensely political in those years (in preparation for her intense life in Italian politics), but says nothing about Bolivian politics. She reiterates her affection for the native population, but always from a distance with a lack of involvement. What she loved about Bolivia was the land, and Indios were included in it: 'I was struck by the beauties of nature. Ultimately, Indios themselves for me were part of the nature of that land. They are the fruit of that land, represent their land; I conceive of them only in those landscapes, in that world.' In a letter home, she wrote: 'The Indio possesses an inborn elegance, a dignified, reserved way, as of a wild and disdainful animal.' She says that native miners were 'content' to live in hovels and huts, sleeping with the animals. Although she does not fail to mention the frequent revolts that took place in the mining camps when miners shouted 'We are thirsty for the White men's blood', there is little effort to investigate how these people attempted to create their own lives and history. Even the interesting discussions of myth, magic, and ceremony (culled from her writings at the time, rather than present memory) are detached and uninvolved, like brilliant travel notes rather than profound experience. Seven years were not enough to overcome the alienness, the estrangement from that distant world she was learning to love.

The book includes both transcribed and edited dialogue between the narrator and the interviewer and editor, Marcella Filippa, and a selection of letters and writings from Levi's personal archive. Sometimes the interaction of these different types of discourse appears a little contrived; often, however, it provides a stimulating perspective, context, and depth.

Alessandro Portelli

Joanne Rappaport, *The Politics of Memory: Native Historical Interpretation in the Colombian Andes*. Cambridge: Cambridge University Press, 1990.

In Latin America, attempts at the totalitarian oppression of native peoples have a long history. In this excellent study, centred on the Paez Indians of the Colombian Andes, Rappaport argues that they have been 'an ethnic group in flux' throughout, but their history has been and is part of what constitutes them as Paez. This changing history is at the interface of the oral and literate.

Key local historians have mediated the literate, Spanish-speaking world of power to the Paez, who equally have had to face it in their own oral realm; history is used to interpret, confront, and try to better the present; as different conditions supervene so do the choices, and representations of the history vary, though seemingly organized through long-lasting images. The *cacique* chiefly office and the *resguardo* territory have changed in scope but remained key political terms for Paez aspiration and debate.

Readers of this *Yearbook* may ask, as Rappaport does, 'whether the Paez have effectively harnessed violence in the modern era, or whether it instead controlled them. Sadly, the reverse is true.' The story is not one which privileges the historical word over forms of domination which are in many ways totalitarian, but it is one which shows how oral history is a form of action. And in Paez, tellers remain significantly creative and activist. It is also an object lesson in how to grasp what accounts mean to their audiences, a persuasive account of native interpretation which other researchers can apply to different cases.

Elizabeth Tonkin

Lawrence Goodwyn, *Breaking the Barrier: The Rise of Solidarity in Poland*. New York: Oxford University Press, 1991.

The breakup of the Communist hold on eastern Europe came as a surprise to most casual observers. In the USA we had been brought up to believe that such an event would not happen in our lifetime. We were especially certain of this after the Solidarity movement in Poland was crushed in 1980. In late August 1989 I visited West Berlin with my family. As we walked along the wall, peeping over into East Berlin, I reminded my family that I had been stationed in Berlin in 1953–4, before the wall was built. I recalled for them the 17 June 1953 uprising, which had been brutally crushed by the Russian occupiers, placing the yoke of oppression even more tightly round the necks of the German people. As we read the wall's graffiti of freedom, I was certain that freedom would continue to lie dormant in the East. Like others I was not conscious that the spirit was loose in Poland and stirring in other countries, and that it would be only a matter of months before the wall, the symbolic yoke of oppression, would crack and crumble.

As historians study post-Second World War Communist oppression, they will attempt to explain why and how the unexpected came about. In their study they will focus on Poland, a country in which the spirit may never have been extinguished. It is in this country that Lawrence Goodwyn, a historian from Duke University, founder of its oral history centre and scholar of the American populist movement, believes one can find the building of a democratic popular movement that was a force for change.

Early reviews of the work have indicated that its thesis will be controversial. Goodwyn maintains that the movement which led to the formation and success of Solidarity sprang out of the factories, that it was the workers and not the intellectuals who gave form and structure to the movement. This thesis serves as a direct challenge to historians, social scientists, and journalists who have done earlier work on the movement, and through whose writings runs, according to Goodwyn, a red thread of class bias. He contends that they were misled by the scarcity of traditional research materials and that they failed to use other research techniques to overcome this shortage.

In a concluding essay, 'Ways of seeing: a critical essay on authorities', Goodwyn separates his research techniques from those of other writers. He attempts to establish that they viewed Polish society 'from afar' (p. 21) by failing to research the place of the workers in the formation and development of the Solidarity movement, and thus holding to 'static' (p. 443) assumptions about movement building. Thus it is no surprise that they provided interpretations and conclusions that were based on false premises.

Unlike the others, Goodwyn understood that it would be necessary to turn to oral testimony to get at the truth. Over a seven-year period, he conducted extensive interviews both with intellectuals and workers. His endnotes give a clear indication that he has used the interviews to support his points, and to aid him in developing his thesis. It is disturbing to find that it is not clear how many interviews were conducted or with whom. Only nine interviews 'were for attribution' (p. 456). He explains that many of the unattributed were with activists who had been imprisoned because of their anti-government activities, or who fled into exile after the brutal crushing of Solidarity in 1980. Some interviewees, fearful of reprisal, refused to make themselves known. Most interviews are cited by giving place of interview, the number assigned, and the date, thus: American Interview No. 16 (New York), 11 July 1985; or Polish Interview No. 2 (Gdansk), 24 April 1983. The notes provide no indication that Goodwyn interviewed Lech Walesa, a central figure in the restructuring of Polish society.

Of greater concern to scholars is the failure to indicate where the interviews and transcripts are deposited, and the terms under which they are available. Hopefully this is an oversight which will be rectified. There can be no doubt that reviews of these interviews are necessary if scholars are going to come to a consensus on the merit of Goodwyn's thesis.

This is a readable and informative work, although sometimes tedious. What gives it most value is that it takes a long-term view of the workers' struggle to form an independent, democratic trade union. While there may be quarrels over parts (if not all) of the Goodwyn thesis, one will be hard put to challenge the idea that democracy existed as a glowing ember in Communist Poland, waiting only for the oxygen of freedom to burst into flame.

John J. Fox

Teresa Toranska (ed.), *'Them': Stalin's Polish Puppets*. New York: Harper and Row, 1988.

The novelist Czeslaw Milosz's analysis of the mental domination accomplished by the State Communist system offered a picture that could hardly be bettered. Equally, Milan Kundera's image of surrealist reality in east European countries was absolutely perfect. And Umberto Eco's two novels had already provided a brilliant study of the ultimate in narrow-mindedness. This book of interviews, edited by Teresa Toranska, achieves the impossible in a different way: it becomes a living, thrilling, and impressive documentary that complements the best literary fiction in its account of the days and the people who for a while thought that they would rule the world forever. Like Eco, she depicts the history of the conspiracy of those who once thought they possessed the secret of life and happiness. A dogma of political and ideological unity in the Communist parties was misunderstood by the keepers of the doctrine, who subsequently committed one error after another, just to retain their well-kept, false secret. Like Kundera, she sheds light on inconsistencies that made grim socialist reality so totally unrealistic. Like Milosz, she focuses on the mechanism that captivated the consciousness of a nation. The only difference is that Toranska's book is not fiction, nor does the author have the same latitude for literary invention.

Toranska has given her book a title that refers to a certain semiotic category of 'Them' (as opposed to 'Us'), a category that has dominated subjective perceptions of Polish social structure in the mid- and late 1970s. 'Them' epitomizes all possible errors and deviations of the former power élite, while 'Us' signifies the new beginnings, a certain political innocence, and the good intentions of the ascending historical generations.

The idea of interviewing people previously prominent in the early Communist regime in Poland was in itself daring. Teresa Toranska started her project back in 1981 with an interview with the interim head of the Communist Party, Edward Ochab, who had inherited the leadership in 1955 from the dictator and Soviet protégé Bierut, and had handed it over to the popularly acclaimed Wladyslaw Gomulka in 1956. In 1981 the Communists were firmly in power, even though waves of workers' protest had already, in practice, questioned the otherwise untouchable idea of State Socialism. If the Communist system had not been dismantled since this project was completed, Toranska's interviews with the founding fathers of Polish Communism might easily have remained what they must have been intended to be—a record illustrating the differences between the *naïveté*, mingled with idealism, of those original élite members from those past times, and the pragmatism, relativism, and a certain cynicism of the Communist élites of the 1980s. When the first interview was published in Poland in the liberal weekly *Polityka*, people read it as if it were a collector's item of Polish history, just to learn about facts that had long been hidden. After all, it had been unusual for top Communist officials to speak freely to independent intellectuals, and most interviews were only granted to licensed Communist journalists.

In the ten years since the first interview, however, Toranska's work, thanks to the unexpected course of history, has become the most crucial source of knowledge about the social consciousness of those who had tried to implement the utopia, and who had failed, without noticing the real cause of their failure. '*Them*' is still a document of the past era, a mine of historical facts previously unknown to students of the darkest period of Polish history. It reveals plots, treasons, and manipulations by foreign agents, and so on. In a sense, it is a testimonial to several socio-historical theories—those of Immanuel Wallerstein, Theda Skocpol, and Charles Tilly—in addition to supplementing the fictional accounts.

It is also an excellent example of the innovative use of the life-story method by a journalist. Unlike the historian or the sociologist, who merely treat their interlocutor as a drawer from which files of facts and data are taken, along with interpretations that are good in themselves, Toranska acts as if the storyteller is no more important than the historian who records the story. She interferes with the interlocutors, and clashes with them, pouring out accusations of lies, nailing them with facts and with her own interpretations, and forcing them to search for better explanations. Toranska's narration is active, frequently aggressive, extremely well informed, and above all reflects a mature vision of history and the processes of social change.

This firsthand information and the personal accounts of the former top officials installed in Poland by the Soviets will undoubtedly be the subject of further study by historians and sociologists in the future. This record of the intentionally selective memories of those who have since passed from view, or sunk into oblivion due to the dramatic course of events, becomes a horrifying picture of the devastating effect that power, wielded without the checks and balances of democracy, may have on its holders. As such, '*Them*' is a memento to all politicians.

Today's reader does not have to be familiar with the myriad of names, dates, and facts from the early history of Polish Communism. When reading '*Them*', one should focus on the dogma that dominated the 1940s and 1950s, and that in fact determined the destinies of millions of people. This dogma, as one of 'Them' reveals, 'was above all dictated by the logic of events which involved all the Socialist countries . . . If they had been different, if we had not been isolated as a bloc, then we would probably have been able to tolerate the coexistence of several parties for a long time . . . ' (p. 289). Toranska, through her searching interviewing, unravels abuses of the credulity of the masses, inconsistencies and lies of the élites, and all the scum that was structurally embedded in State Socialism.

In the meantime, the method and style initiated ten years ago has now found a central place in the history of the past regime. Two volumes of the life story of Edward Gierek, the leader who was toppled by Solidarity in 1980, have scooped the Polish market, and are now being translated into several languages. Similarly, the book-long interview with Wlodzimierz Sokorski, another sinister figure from the past decades, has recently been published. One would also like to hear comrades Erich Honnecker, Thodor Zhivkov, Janos Kadar, and Gustav Husak talking to interviewers as equally skilled and intelligent as Teresa Toranska.

Finally, following the collection of 'Them' talking about their past, one would like to see another like it, where 'Us' would be talking as well.

Bronislaw Misztal

Alain Brossat, Sonia Combe, Jean-Yves Potel, and Jean-Charles Szurek (eds.), *A l'Est la mémoire retrouvée* (To the east: Memory Rediscovered); preface by Jacques Le Goff. Paris: La Découverte,

The theme of this amazing volume is the no less amazing process of recovering a collective memory in the countries formerly ruled by Communist regimes. This process has just begun. It is by no means a natural process; on the contrary, it takes incredible efforts of inquiry into past events, putting aside deeply ingrained and endlessly repeated official statements, which prove to be downright lies, and cracking the nuts of police secrecy to begin to understand how all this could ever have happened and, for instance, to find out what exactly happened to one's own parents and grandparents.

The twenty-four authors are mostly young historians from the Soviet Union, Poland, Romania, and East Germany; or French and Italian scholars, who are specialists in Soviet block countries and are spending part of their own life there. Their papers each focus on a particular aspect, locally observed and concretely described, of the overall 'struggle for reconstructing the past', and hence for overturning both a collective and an individual identity, and historical consciousness.

For instance, Maria Ferretti describes how thousands of adults in the Soviet Union are presently trying to gain access to the archives of the KGB to learn at long last *why* their parents were sentenced, where they were sent to, and how and when they finally died. When one realizes that not only millions of Soviet people were sent to the Gulag—a large proportion of whom died there—but that also whole populations were deported from one end of the Union to the other—and died of hunger, disease, or sheer exhaustion, without leaving any trace of their passage through this world except perhaps one child—one realizes how wide the gap is between the desire for concrete historical information and actual knowledge. Glasnost has not yet been extended to *this* dark basement of Soviet history; that will only be so if people eventually force their way into it.

Oral sources may play a key role in the process, as they did when Solzhenitsyn began collecting evidence for his Gulag book. But the grip of totalitarianism extended far beyond the sphere of public discourse, where its control was total: it reached into the sphere of private life by preventing parents from telling their children what they had experienced, lest they should end up being denounced for anti-Socialist propaganda.

Some papers tell us about some of the forms that official authorities gave to the collective memory they were fabricating, by showing the tools used in the cult of Lenin, the East German commemorations as described by Sonia Combe, the ways national museums presented national history in, for example, Romania, or the grotesque museum of the police in Prague, wittily narrated by Véronique Soulé. Other papers describe some of the collective struggles to gain access to the past (Faulkner's quote 'the past is never dead, it is not even passed' applies well to the situation), for instance in Evgeni Kojokine's account of the Tartars deported from the Crimea; or struggles by Balts or Jews. Drawing on all these concrete studies, the editors and also Jacques Le Goff, a prominent French historian, develop new lines of thought about the complex relations between history, collective memory, historical consciousness, and identity. Le Goff, who has written extensively on the topic *Mémoire et histoire*, says: 'this volume is essential not only to understand what is happening in eastern Europe, but also to understand our epoch, and the whole relationship between memory and history'.

Indeed, reading this series of papers one realizes how much we take for granted—those of us who are fortunate enough to live in democratic regimes—such things as national identity, historical consciousness, and collective (and family) memory. All these are part of our mental landscape; they seem to have always been there, and they are grounded in actual historical events whose reality and meaning are shared by practically all citizens. We tend to forget that identity, consciousness, memory itself are all outcomes of very specific social conditions; they are in fact social phenomena, outcomes of ever-active processes of the construction of meaning through contradictory debates that are only possible where freedom of expression, initiatives by individual citizens, and institutions allowing free-thinking are consolidated. By contrast, this volume gives many examples of the *struggle for memory* that is presently taking place alongside the struggle for history in post-totalitarian countries.

Daniel Bertaux

BOOKS FOR REVIEW and review correspondence should be sent to Selma Leydesdorff, Belle van Zuylen Ondersoeks Instituut, Oudezijds Achterburgwal 185, 1012 DK Amsterdam, The Netherlands.

OFFERS OF PAPERS and editorial correspondence to Paul Thompson, 18 Lonsdale Road, Oxford OX2 7EW, England.

SUBSCRIPTION INQUIRIES to Oxford University Press, Walton St. Oxford OX2 6DP, England.